SYMMETRY

BY HERMANN WEYL

PRINCETON UNIVERSITY PRESS,

PRINCETON, NEW JERSEY

LCC 52-5830
ISBN 0-691-08045-3
ISBN 0-691-02374-3 pbk.

Printed in the United States of America
by Princeton University Press,
Princeton, New Jersey

First Princeton Paperback printing, 1982

PREFACE

AND BIBLIOGRAPHICAL REMARKS

STARTING from the somewhat vague notion of symmetry = harmony of proportions, these four lectures gradually develop first the geometric concept of symmetry in its several forms, as bilateral, translatory, rotational, ornamental and crystallographic symmetry, etc., and finally rise to the general idea underlying all these special forms, namely that of invariance of a configuration of elements under a group of automorphic transformations. I aim at two things: on the one hand to display the great variety of applications of the principle of symmetry in the arts, in inorganic and organic nature, on the other hand to clarify step by step the philosophico-mathematical significance of the idea of symmetry. The latter purpose makes it necessary to confront the notions and theories of symmetry and relativity, while numerous illustrations supporting the text help to accomplish the former.

As readers of this book I had a wider circle in mind than that of learned specialists. It does not shun mathematics (that would defeat its purpose), but detailed treatment of most of the problems it deals with, in particular complete mathematical treatment, is beyond its scope. To the lectures, which reproduce in slightly modified version the Louis Clark Vanuxem Lectures given by the author at Princeton University in February 1951, two appendices containing mathematical proofs have been added.

Other books in the field, as for instance F. M. Jaeger's classical *Lectures on the principle of symmetry and its applications in natural science* (Amsterdam and London, 1917), or the much smaller and more recent booklet by Jacque Nicolle, *La symétrie et ses applications* (Paris, Albin Michel, 1950) cover only part of the material, though in a more detailed fashion. Symmetry is but a side-issue in D'Arcy Thompson's magnificent work *On growth and form* (New edition, Cambridge, Engl., and New York, 1948). Andreas Speiser's *Theorie der Gruppen von endlicher Ordnung* (3. Aufl. Berlin, 1937) and other publications by the same author are important for the synopsis of the aesthetic and mathematical aspects of the subject. Jay Hambidge's *Dynamic symmetry* (Yale University Press, 1920) has little more than the name in common with the present book. Its closest relative is perhaps the July 1949 number on symmetry of the German periodical *Studium Generale* (Vol. 2, pp. 203–278: quoted as *Studium Generale*).

A complete list of sources for the illustrations is to be found at the end of the book.

To the Princeton University Press and its editors I wish to express warm thanks for the inward and outward care they have lavished on this little volume; to the authorities of Princeton University no less sincere thanks for the opportunity they gave me to deliver this swan song on the eve of my retirement from the Institute for Advanced Study.

HERMANN WEYL

Zurich
December 1951

CONTENTS

BILATERAL SYMMETRY

BILATERAL SYMMETRY

IF I AM NOT MISTAKEN the word *symmetry* is
used in our everyday language in two mean-
ings. In the one sense symmetric means
something like well-proportioned, well-bal-
anced, and symmetry denotes that sort of con-
cordance of several parts by which they inte-
grate into a whole. *Beauty* is bound up with
symmetry. Thus Polykleitos, who wrote a
book on proportion and whom the ancients
praised for the harmonious perfection of his
sculptures, uses the word, and Dürer follows
him in setting down a canon of proportions
for the human figure.[1] In this sense the idea
is by no means restricted to spatial objects;
the synonym "harmony" points more toward
its acoustical and musical than its geometric
applications. *Ebenmass* is a good German
equivalent for the Greek symmetry; for like
this it carries also the connotation of "middle

[1] Dürer, *Vier Bücher von menschlicher Proportion*, 1528.
To be exact, Dürer himself does not use the word
symmetry, but the "authorized" Latin translation by
his friend Joachim Camerarius (1532) bears the title
De symmetria partium. To Polykleitos the statement is
ascribed (περὶ βελοποιϊκῶν, ιν, 2) that "the employment
of a great many numbers would almost engender
correctness in sculpture." See also Herbert Senk, Au
sujet de l'expression συμμετρία dans Diodore ι, 98,
5–9, in *Chronique d'Egypte 26* (1951), pp. 63–66.
Vitruvius defines: "Symmetry results from proportion . . . Proportion is the commensuration of the
various constituent parts with the whole." For a
more elaborate modern attempt in the same direction
see George David Birkhoff, *Aesthetic measure*, Cambridge, Mass., Harvard University Press. 1933, and the
lectures by the same author on "A mathematical
theory of aesthetics and its applications to poetry
and music," *Rice Institute Pamphlet*, Vol. 19 (July,
1932), pp. 189–342.

measure," the mean toward which the virtuous should strive in their actions according to Aristotle's Nicomachean Ethics, and which Galen in *De temperamentis* describes as that state of mind which is equally removed from both extremes: σύμμετρον ὅπερ ἑκατέρου τῶν ἄκρων ἀπέχει.

The image of the balance provides a natural link to the second sense in which the word symmetry is used in modern times: *bilateral symmetry,* the symmetry of left and right, which is so conspicuous in the structure of the higher animals, especially the human body. Now this bilateral symmetry is a strictly geometric and, in contrast to the vague notion of symmetry discussed before, an absolutely precise concept. A body, a spatial configuration, is symmetric with respect to a given plane E if it is carried into itself by reflection in E. Take any line l perpendicular to E and any point p on l: there exists one and only one point p' on l which has the same distance from E but lies on the other side. The point p' coincides with p only if p is on E. Reflection in E is that mapping

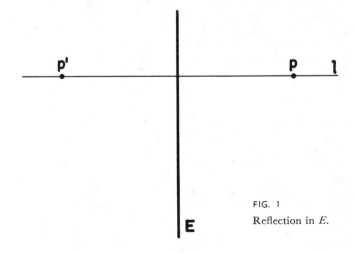

FIG. 1

Reflection in E.

of space upon itself, $S: p \rightarrow p'$, that carries the arbitrary point p into this its mirror image p' with respect to E. A mapping is defined whenever a rule is established by which every point p is associated with an image p'. Another example: a rotation around a perpendicular axis, say by 30°, carries each point p of space into a point p' and thus defines a mapping. A figure has rotational symmetry around an axis l if it is carried into itself by all rotations around l. Bilateral symmetry appears thus as the first case of a geometric concept of symmetry that refers to such operations as reflections or rotations. Because of their complete rotational symmetry, the circle in the plane, the sphere in space were considered by the Pythagoreans the most perfect geometric figures, and Aristotle ascribed spherical shape to the celestial bodies because any other would detract from their heavenly perfection. It is in this tradition that a modern poet[2] addresses the Divine Being as "Thou great symmetry":

> *God, Thou great symmetry,*
> *Who put a biting lust in me*
> *From whence my sorrows spring,*
> *For all the frittered days*
> *That I have spent in shapeless ways*
> *Give me one perfect thing.*

Symmetry, as wide or as narrow as you may define its meaning, is one idea by which man through the ages has tried to comprehend and create order, beauty, and perfection.

The course these lectures will take is as follows. First I will discuss bilateral symmetry in some detail and its role in art as

[2] Anna Wickham, "Envoi," from *The contemplative quarry*, Harcourt, Brace and Co., 1921.

well as organic and inorganic nature. Then we shall generalize this concept gradually, in the direction indicated by our example of rotational symmetry, first staying within the confines of geometry, but then going beyond these limits through the process of mathematical abstraction along a road that will finally lead us to a mathematical idea of great generality, the Platonic idea as it were behind all the special appearances and applications of symmetry. To a certain degree this scheme is typical for all theoretic knowledge: We begin with some general but vague principle (symmetry in the first sense), then find an important case where we can give that notion a concrete precise meaning (bilateral symmetry), and from that case we gradually rise again to generality, guided more by mathematical construction and abstraction than by the mirages of philosophy; and if we are lucky we end up with an idea no less universal than the one from which we started. Gone may be much of its emotional appeal, but it has the same or even greater unifying power in the realm of thought and is exact instead of vague.

I open the discussion on bilateral symmetry by using this noble Greek sculpture from the fourth century B.C., the statue of a praying boy (Fig. 2), to let you feel as in a symbol the great significance of this type of symmetry both for life and art. One may ask whether the aesthetic value of symmetry depends on its vital value: Did the artist discover the symmetry with which nature according to some inherent law has endowed its creatures, and then copied and perfected what nature presented but in imperfect realizations; or has the aesthetic value of symmetry an independent source? I am in-

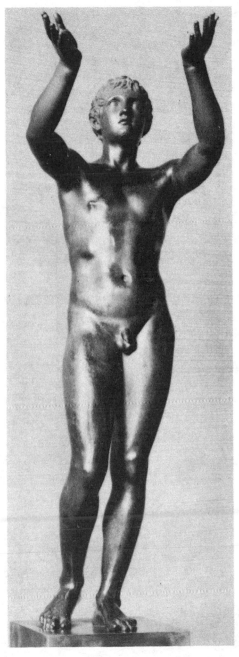

FIG. 2

clined to think with Plato that the mathematical idea is the common origin of both: the mathematical laws governing nature are the origin of symmetry in nature, the intuitive realization of the idea in the creative artist's mind its origin in art; although I am ready to admit that in the arts the fact of the bilateral symmetry of the human body in its outward appearance has acted as an additional stimulus.

Of all ancient peoples the Sumerians seem to have been particularly fond of strict bilateral or heraldic symmetry. A typical design on the famous silver vase of King Entemena, who ruled in the city of Lagash

FIG. 3

around 2700 B.C., shows a lion-headed eagle
with spread wings *en face*, each of whose claws
grips a stag in side view, which in its turn is
frontally attacked by a lion (the stags in the
upper design are replaced by goats in the
lower) (Fig. 3). Extension of the exact symmetry of the eagle to the other beasts obviously enforces their duplication. Not much
later the eagle is given two heads facing in
either direction, the formal principle of symmetry thus completely overwhelming the
imitative principle of truth to nature. This
heraldic design can then be followed to
Persia, Syria, later to Byzantium, and anyone
who lived before the First World War will
remember the double-headed eagle in the
coats-of-arms of Czarist Russia and the
Austro-Hungarian monarchy.

Look now at this Sumerian picture (Fig. 4).
The two eagle-headed men are nearly but
not quite symmetric; why not? In plane
geometry reflection in a vertical line *l* can
also be brought about by rotating the plane
in space around the axis *l* by 180°. If you
look at their arms you would say these two

monsters arise from each other by such rotation; the overlappings depicting their position in space prevent the plane picture from having bilateral symmetry. Yet the artist aimed at that symmetry by giving both figures a half turn toward the observer and also by the arrangement of feet and wings: the drooping wing is the right one in the left figure, the left one in the right figure.

FIG. 5

The designs on the cylindrical Babylonian seal stones are frequently ruled by heraldic symmetry. I remember seeing in the collection of my former colleague, the late Ernst Herzfeld, samples where for symmetry's sake not the head, but the lower bull-shaped part of a god's body, rendered in profile, was doubled and given four instead of two hind

FIG. 6

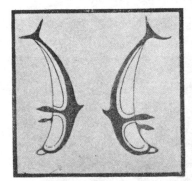

FIG. 7

legs. In Christian times one may see an analogy in certain representations of the Eucharist as on this Byzantine paten (Fig. 5), where two symmetric Christs are facing the disciples. But here symmetry is not complete and has clearly more than formal significance, for Christ on one side breaks the bread, on the other pours the wine.

Between Sumeria and Byzantium let me insert Persia: These enameled sphinxes (Fig. 6) are from Darius' palace in Susa built in

11

the days of Marathon. Crossing the Aegean
we find these floor patterns (Fig. 7) at the
Megaron in Tiryns, late helladic about 1200
B.C. Who believes strongly in historic continuity and dependence will trace the grace-
ful designs of marine life, dolphin and
octopus, back to the Minoan culture of Crete,
the heraldic symmetry to oriental, in the
last instance Sumerian, influence. Skipping
thousands of years we still see the same influences at work in this plaque (Fig. 8) from the
altar enclosure in the dome of Torcello,
Italy, eleventh century A.D. The peacocks
drinking from a pine well among vine leaves
are an ancient Christian symbol of immor-
tality, the structural heraldic symmetry is
oriental.

FIG. 8

FIG. 9

For in contrast to the orient, occidental art, like life itself, is inclined to mitigate, to loosen, to modify, even to break strict symmetry. But seldom is asymmetry merely the absence of symmetry. Even in asymmetric designs one feels symmetry as the norm from which one deviates under the influence of forces of non-formal character. I think the riders from the famous Etruscan Tomb of the Triclinium at Corneto (Fig. 9) provide a good example. I have already mentioned representations of the Eucharist with Christ duplicated handing out bread and wine. The central group, Mary flanked by two angels, in this mosaic of the Lord's Ascension (Fig. 10) in the cathedral at Monreale, Sicily (twelfth century), has almost perfect symmetry. [The band ornament's above and below the mosaic will demand our attention in the second lecture.] The principle of symmetry is somewhat less strictly observed in an earlier mosaic from San Apollinare in

FIG. 10

Ravenna (Fig. 11), showing Christ surrounded by an angelic guard of honor. For
instance Mary in the Monreale mosaic raises
both hands symmetrically, in the *orans* ges-
ture; here only the right hands are raised.
Asymmetry has made further inroads in the
next picture (Fig. 12), a Byzantine relief
ikon from San Marco, Venice. It is a Deësis,
and, of course, the two figures praying for
mercy as the Lord is about to pronounce the
last judgment cannot be mirror images of
each other; for to the right stands his Virgin

14

FIG. 11

Mother, to the left John the Baptist. You may also think of Mary and John the Evangelist on both sides of the cross in crucifixions as examples of broken symmetry.

FIG. 12

Clearly we touch ground here where the precise geometric notion of bilateral symmetry begins to dissolve into the vague notion of *Ausgewogenheit*, balanced design with which we started. "Symmetry," says Dagobert Frey in an article *On the Problem of Symmetry in Art*,[3] "signifies rest and binding, asymmetry motion and loosening, the one order and law, the other arbitrariness and accident, the one formal rigidity and constraint, the other life, play and freedom." Wherever God or Christ are represented as symbols for everlasting truth or justice they are given in the symmetric frontal view, not in profile. Probably for similar reasons public buildings and houses of worship, whether they are Greek temples or Christian basilicas and cathedrals, are bilaterally symmetric. It is, however, true that not infrequently the two towers of Gothic cathedrals are different, as for instance in Chartres. But in practically every case this seems to be due to the history of the cathedral, namely to the fact that the towers were built in different periods. It is understandable that a later time was no longer satisfied with the design of an earlier period; hence one may speak here of historic asymmetry. Mirror images occur where there is a mirror, be it a lake reflecting a landscape or a glass mirror into which a woman looks. Nature as well as painters make use of this motif. I trust, examples will easily come to your mind. The one most familiar to me, because I look at it in my study every day, is Hodler's *Lake of Silvaplana*.

While we are about to turn from art to nature, let us tarry a few minutes and first consider what one may call the *mathematical philosophy of left and right*. To the scientific

[3] Studium Generale, p. 276.

mind there is no inner difference, no polarity between left and right, as there is for instance in the contrast of male and female, or of the anterior and posterior ends of an animal. It requires an arbitrary act of choice to determine what is left and what is right. But after it is made for one body it is determined for every body. I must try to make this a little clearer. In space the distinction of left and right concerns the orientation of a screw. If you speak of turning left you mean that the sense in which you turn combined with the upward direction from foot to head of your body forms a left screw. The daily rotation of the earth together with the direction of its axis from South to North Pole is a left screw, it is a right screw if you give the axis the opposite direction. There are certain crystalline substances called optically active which betray the inner asymmetry of their constitution by turning the polarization plane of polarized light sent through them either to the left or to the right; by this, of course, we mean that the sense in which the plane rotates while the light travels in a definite direction, combined with that direction, forms a left screw (or a right one, as the case may be). Hence when we said above and now repeat in a terminology due to Leibniz, that left and right are *indiscernible*, we want to express that the inner structure of space does not permit us, except by arbitrary choice, to distinguish a left from a right screw.

I wish to make this fundamental notion still more precise, for on it depends the entire theory of relativity, which is but another aspect of symmetry. According to Euclid one can describe the structure of space by a number of basic relations between points, such as *ABC* lie on a straight line, *ABCD* lie in a plane, AB is congruent CD. Perhaps the best way of describing the structure of space is the one Helmholtz adopted: by the single notion of *congruence* of figures. A mapping S of space associates with every point p a point $p':p \rightarrow p'$. A pair of mappings S, $S':p \rightarrow p'$, $p' \rightarrow p$, of which the one is the inverse of the other, so that if S carries p into p' then S'

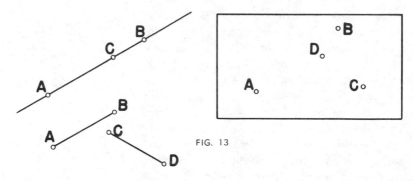

FIG. 13

carries p' back into p and vice versa, is spoken of as a pair of one-to-one mappings or *transformations*. A transformation which preserves the structure of space—and if we define this structure in the Helmholtz way, that would mean that it carries any two congruent figures into two congruent ones—is called an *automorphism* by the mathematicians. Leibniz recognized that this is the idea underlying the geometric concept of similarity. An automorphism carries a figure into one that in Leibniz' words is "indiscernible from it if each of the two figures is considered by itself." What we mean then by stating that left and right are of the same essence is the fact that *reflection in a plane is an automorphism*.

Space as such is studied by geometry. But space is also the medium of all physical occurrences. The structure of the physical world is revealed by the general laws of nature. They are formulated in terms of certain basic quantities which are functions in space and time. We would conclude that the physical structure of space "contains a screw," to use a suggestive figure of speech, if

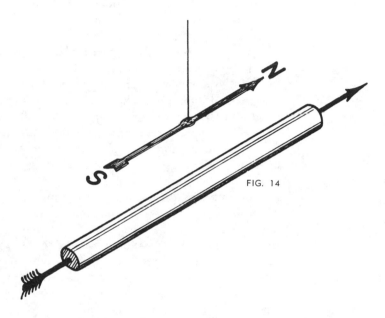

FIG. 14

these laws were not invariant throughout with respect to reflection. Ernst Mach tells of the intellectual shock he received when he learned as a boy that a magnetic needle is deflected in a certain sense, to the left or to the right, if suspended parallel to a wire through which an electric current is sent in a definite direction (Fig. 14). Since the whole geometric and physical configuration, including the electric current and the south and north poles of the magnetic needle, to all appearances, are symmetric with respect to the plane E laid through the wire and the needle, the needle should react like Buridan's ass between

19

equal bundles of hay and refuse to decide
between left and right, just as scales of equal
arms with equal weights neither go down on
their left nor on their right side but stay
horizontal. But appearances are sometimes
deceptive. Young Mach's dilemma was the
result of a too hasty assumption concerning
the effect of reflection in E on the electric
current and the positive and negative magnetic poles of the needle: while we know a
priori how geometric entities fare under
reflection, we have to learn from nature how
the physical quantities behave. And this is
what we find: under reflection in the plane E
the electric current preserves its direction, but
the magnetic south and north poles are interchanged. Of course this way out, which re-
establishes the equivalence of left and right, is
possible only because of the essential equality
of positive and negative magnetism. All
doubts were dispelled when one found that
the magnetism of the needle has its origin in
molecular electric currents circulating around
the needle's direction; it is clear that under
reflection in the plane E such currents
change the sense in which they flow.

The net result is that in all physics nothing
has shown up indicating an intrinsic difference of left and right. Just as all points and
all directions in space are equivalent, so are
left and right. Position, direction, left and
right are *relative* concepts. In language tinged
with theology this issue of relativity was discussed at great length in a famous controversy
between Leibniz and Clarke, the latter a
clergyman acting as the spokesman for
Newton.[4] Newton with his belief in absolute

[4] See G. W. Leibniz, *Philosophische Schriften*, ed.
Gerhardt (Berlin 1875 seq.), VII, pp. 352–440, in
particular Leibniz' third letter, §5.

space and time considers motion a proof of the creation of the world out of God's arbitrary will, for otherwise it would be inexplicable why matter moves in this rather than in any other direction. Leibniz is loath to burden God with such decisions as lack "sufficient reason." Says he, "Under the assumption that space be something in itself it is impossible to give a reason why God should have put the bodies (without tampering with their mutual distances and relative positions) just at this particular place and not somewhere else; for instance, why He should not have arranged everything in the opposite order by turning East and West about. If, on the other hand, space is nothing more than the spatial order and relation of things then the two states supposed above, the actual one and its transposition, are in no way different from each other . . . and therefore it is a quite inadmissible question to ask why one state was preferred to the other." By pondering the problem of left and right Kant was first led to his conception of space and time as forms of intuition.[5] Kant's opinion seems to have been this: If the first creative act of God had been the forming of a left hand then this hand, even at the time when it could be compared to nothing else, had the distinctive character of left, which can only intuitively but never conceptually be apprehended. Leibniz contradicts: According to him it would have made no difference if God had created a "right" hand first rather than a "left" one. One must follow the world's creation a step further before a difference can appear. Had God, rather than

---

[5] Besides his "Kritik der reinen Vernunft" see especially §13 of the *Prolegomena zu einer jeden künftigen Metaphysik.* . .

making first a left and then a right hand, started with a right hand and then formed another right hand, He would have changed the plan of the universe not in the first but in the second act, by bringing forth a hand which was equally rather than oppositely oriented to the first-created specimen.

Scientific thinking sides with Leibniz. Mythical thinking has always taken the contrary view as is evinced by its usage of right and left as symbols for such polar opposites as good and evil. You need only think of the double meaning of the word *right* itself. In this detail from Michelangelo's famous *Creation of Adam* from the Sistine Ceiling (Fig. 15) God's right hand, on the right, touches life into Adam's left.

People shake right hands. *Sinister* is the Latin word for left, and heraldry still speaks of the left side of the shield as its sinister side. But *sinistrum* is at the same time that which is evil, and in common English only this figurative meaning of the Latin word survives.[6] Of the two malefactors who were crucified with Christ, the one who goes with Him to paradise is on His right. St. Matthew, Chapter 25, describes the last judgment as follows: "And he shall set the sheep on his right hand but the goats on the left. Then shall the King say unto them on his right hand, Come ye, blessed of my Father, inherit the Kingdom prepared for you from the foundation of the world. . . . Then he shall say also unto them on the left hand, Depart from me, ye cursed, into everlasting fire, prepared for the devil and his angels."

[6] I am not unaware of the strange fact that as a *terminus technicus* in the language of the Roman augurs *sinistrum* had just the opposite meaning of propitious.

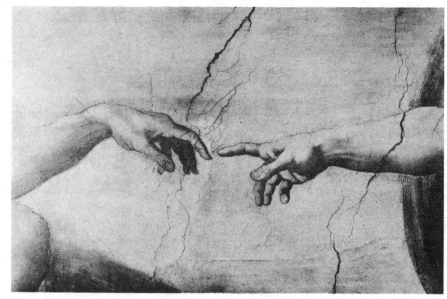

FIG. 15

I remember a lecture Heinrich Wölfflin once delivered in Zurich on "Right and left in paintings"; together with an article on "The problem of inversion (Umkehrung) in Raphael's tapistry cartoons," you now find it printed in abbreviated form in his *Gedanken zur Kunstgeschichte*, 1941. By a number of examples, as Raphael's *Sistine Madonna* and Rembrandt's etching *Landscape with the three trees*, Wölfflin tries to show that right in painting has another *Stimmungswert* than left. Practically all methods of reproduction interchange left and right, and it seems that former times were much less sensitive than we are toward such inversion. (Even Rembrandt did not hesitate to bring his Descent from the Cross as a converse etching upon the market.) Considering that we do a lot more reading than the people, say, of the sixteenth century, this suggests the hypothesis that the difference pointed out by Wölfflin is connected with our habit of reading from left to right. As far as I remember, he himself rejected this as well as a number of other psychological explanations put forward in the discussion after his lecture. The printed text concludes with the remark that the problem "obviously has deep roots, roots which reach down to the very foundations of our sensuous nature." I for my part am disinclined to take the matter that seriously.[7]

In science the belief in the equivalence of left and right has been upheld even in the face of certain biological facts presently to be mentioned which seem to suggest their inequivalence even more strongly than does the deviation of the magnetic needle which shocked young Mach. The same problem of equivalence arises with respect to *past and future*, which are interchanged by inverting the direction of time, and with respect to *positive and negative electricity*. In these cases, especially in the second, it is perhaps clearer than for the pair left-right that a priori evidence is not sufficient to settle the question; the empirical facts have to be consulted. To be sure, the role which past and future play in our consciousness would indicate their intrinsic difference—the past knowable and unchangeable, the future unknown and still alterable by decisions taken now—and one would expect that this difference has its basis in the physical laws of nature. But those laws of which we can boast a reasonably certain knowledge are invariant with respect

[7] Cf. also A. Faistauer, "Links und rechts im Bilde," Amicis, *Jahrbuch der österreichischen Galerie*, 1926, p. 77; Julius v. Schlosser, "Intorno alla lettura dei quadri," *Critica 28*, 1930, p. 72; Paul Oppé, "Right and left in Raphael's cartoons," *Journal of the Warburg and Courtauld Institutes* 7, 1944, p. 82.

to the inversion of time as they are with respect to the interchange of left and right. Leibniz made it clear that the temporal modi past and future refer to the *causal structure* of the world. Even if it is true that the exact "wave laws" formulated by quantum physics are not altered by letting time flow backward, the metaphysical idea of causation, and with it the one way character of time, may enter physics through the statistical interpretation of those laws in terms of probability and particles. Our present physical knowledge leaves us even more uncertain about the equivalence or non-equivalence of positive and negative electricity. It seems difficult to devise physical laws in which they are not intrinsically alike; but the negative counterpart of the positively charged proton still remains to be discovered.

This half-philosophical excursion was needed as a background for the discussion of the left-right symmetry in nature; we had to understand that the general organization of nature possesses that symmetry. But one will not expect that any special object of nature shows it to perfection. Even so, it is surprising to what extent it prevails. There must be a reason for this, and it is not far to seek: a state of equilibrium is likely to be symmetric. More precisely, under conditions which determine a unique state of equilibrium the symmetry of the conditions must carry over to the state of equilibrium. Therefore tennis balls and stars are spheres; the earth would be a sphere too if it did not rotate around an axis. The rotation flattens it at the poles but the rotational or cylindrical symmetry around its axis is preserved. The feature that needs explanation is, therefore, not the rotational symmetry of its shape but

25

the deviations from this symmetry as exhibited by the irregular distribution of land and water and by the minute crinkles of mountains on its surface. It is for such reasons that in his monograph on the left-right problem in zoology Wilhelm Ludwig says hardly a word about the origin of the bilateral symmetry prevailing in the animal kingdom from the echinoderms upward, but in great detail discusses all sorts of secondary asymmetries superimposed upon the symmetrical ground plan.[8] I quote: "The human body like that of the other vertebrates is basically built bilateral-symmetrically. All asymmetries occurring are of secondary character, and the more important ones affecting the inner organs are chiefly conditioned by the necessity for the intestinal tube to increase its surface out of proportion to the growth of the body, which lengthening led to an asymmetric folding and rolling-up. And in the course of phylogenetic evolution these first asymmetries concerning the intestinal system with its appendant organs brought about asymmetries in other organ systems." It is well known that the heart of mammals is an asymmetric screw, as shown by the schematic drawing of Fig. 16.

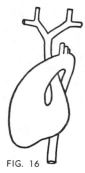

FIG. 16

If nature were all lawfulness then every phenomenon would share the full symmetry of the universal laws of nature as formulated by the theory of relativity. The mere fact that this is not so proves that *contingency* is an essential feature of the world. Clarke in his controversy with Leibniz admitted the latter's principle of sufficient reason but added that the sufficient reason often lies in the mere will of God. I think, here Leibniz the rationalist

---

[8] W. Ludwig, *Rechts-links-Problem im Tierreich und beim Menschen*, Berlin 1932.

is definitely wrong and Clarke on the right track. But it would have been more sincere to deny the principle of sufficient reason altogether instead of making God responsible for all that is unreason in the world. On the other hand Leibniz was right against Newton and Clarke with his insight into the principle of relativity. The truth as we see it today is this: The laws of nature do not determine uniquely the one world that actually exists, not even if one concedes that two worlds arising from each other by an automorphic transformation, i.e., by a transformation which preserves the universal laws of nature, are to be considered the same world.

If for a lump of matter the overall symmetry inherent in the laws of nature is limited by nothing but the accident of its position P then it will assume the form of a sphere around the center P. Thus the lowest forms of animals, small creatures suspended in water, are more or less spherical. For forms fixed to the bottom of the ocean the direction of gravity is an important factor, narrowing the set of symmetry operations from all rotations around the center P to all rotations about an axis. But for animals capable of self-motion in water, air, or on land both the postero-anterior direction in which their body moves and the direction of gravity are of decisive influence. After determination of the antero-posterior, the dorso-ventral, and thereby of the left-right axes, only the distinction between left and right remains arbitrary, and at this stage no higher symmetry than the bilateral type can be expected. Factors in the phylogenetic evolution that tend to introduce inheritable differences between left and right are likely to be held in check by the advantage an animal derives from the bilateral formation of its organs of motion, cilia or muscles and limbs: in case of their asymmetric development a screw-wise instead of a straight-forward motion would naturally result. This may help to explain why our limbs obey the law of symmetry more strictly than our inner organs. Aristophanes in Plato's *Symposium* tells a different story of how the transition from spherical to bilateral symmetry came about. Originally, he says, man was round, his back and sides forming a circle. To humble their pride and might Zeus cut them into two and had Apollo turn their faces and genitals around; and Zeus has threatened, "If they continue insolent I will split them again and they shall hop around on a single leg."

The most striking examples of symmetry in the inorganic world are the crystals. The gaseous and the crystalline are two clear-cut states of matter which physics finds relatively easy to explain; the states in between these two extremes, like the fluid and the plastic states, are somewhat less amenable to theory. In the gaseous state molecules move freely around in space with mutually independent random positions and velocities. In the crystalline state atoms oscillate about positions of equilibrium as if they were tied to them by elastic strings. These positions of equilibrium form a fixed regular configuration in space. What we mean by regular and how the visible symmetry of crystals derives from the regular atomic arrangement will be explained in a subsequent lecture. While most of the thirty-two geometrically possible systems of crystal symmetry involve bilateral symmetry, not all of them do. Where it is not involved we have the possibility of so-called enantiomorph crystals which exist in a laevo- and dextro-form, each form being a mirror image of the other, like left and right hands. A substance which is optically active, i.e., turns the plane of polarized light either left or right, can be expected to crystallize in such asymmetric forms. If the laevo-form exists in nature one would assume that the dextro-form exists likewise, and that in the average both occur with equal frequencies. In 1848 Pasteur made the discovery that when the sodium ammonium salt of optically inactive racemic acid was recrystallized from an aqueous solution at a lower temperature the deposit consisted of two kinds of tiny crystals which were mirror images of each other. They were carefully separated, and the acids set free from the one and the other proved to have the same chemical composition as the racemic acid, but one was optically laevo-active, the other dextro-active. The latter was found to be identical with the tartaric acid present in fermenting grapes, the other had never before been observed in nature. "Seldom," says F. M. Jaeger in his lectures *On the principle of symmetry and its applications in natural science*, "has a scientific discovery had such far-reaching consequences as this one had."

Quite obviously some accidents hard to control decide whether at a spot of the solution a laevo- or dextro-crystal comes into being; and thus in agreement with the symmetric and optically inactive character of the solution as a whole and with the law of chance the amounts of substance deposited in the one and the other form at any moment of the process of crystallization are equal or very nearly equal. On the other hand nature, in giving us the wonderful gift of grapes so much enjoyed by Noah, produced only one of the forms, and it remained for Pasteur to produce the other! This is strange indeed. It is a fact that most of the numerous carbonic compounds occur in nature in one, either the laevo- or the dextro-form only. The sense in which a snail's shell winds is an inheritable character founded in its genetic constitution, as is the "left heart" and the winding of the intestinal duct in the species *Homo sapiens*. This does not exclude that in-inversions occur, e.g. *situs inversus* of the intestines of man occurs with a frequency of about 0.02 per cent; we shall come back to that later! Also the deeper chemical constitution of our human body shows that we have a screw, a screw that is turning the same way in every one of us. Thus our body contains the dextro-rotatory form of glucose and laevo-rotatory form of fructose. A horrid manifestation of this genotypical asymmetry is a metabolic disease called phenylketonuria, leading to insanity, that man contracts when a small quantity of laevo-phenylalanine is added to his food, while the dextro-form has no such disastrous effects. To the asymmetric chemical constitution of living organisms one must attribute the success of Pasteur's method of isolating the laevo- and dextro-forms of substances by means of the enzymatic action of bacteria, moulds, yeasts, and the like. Thus he found that an originally inactive solution of some racemate became gradually laevo-rotatory if *Penicillium glaucum* was grown in it. Clearly the organism selected for its nutriment that form of the tartaric acid molecule which best suited its own asymmetric chemical constitution. The image of lock and key has been used to illustrate this specificity of the action of organisms.

In view of the facts mentioned and in view

of the failure of all attempts to "activate" by mere chemical means optically inactive material,[9] it is understandable that Pasteur clung to the opinion that the production of single optically active compounds was the very prerogative of life. In 1860 he wrote, "This is perhaps the only well-marked line of demarkation that can at present be drawn between the chemistry of dead and living matter." Pasteur tried to explain his very first experiment where racemic acid was transformed by recrystallization into a mixture of laevo- and dextro-tartaric acid by the action of bacteria in the atmosphere on his neutral solution. It is quite certain today that he was wrong; the sober physical explanation lies in the fact that at lower temperature a mixture of the two oppositely active tartaric forms is more stable than the inactive racemic form. If there is a difference in principle between life and death it does not lie in the chemistry of the material substratum; this has been fairly certain ever since Wöhler in 1828 synthesized urea from purely mineral material. But even as late as 1898 F. R. Japp in a famous lecture on "Stereochemistry and Vitalism" before the British Association upheld Pasteur's view in the modified form: "Only the living organisms, or the living intelligence with its conception of symmetry can produce this result (i.e. asymmetric compounds)." Does he really mean that it is Pasteur's intelligence that, by devising the experiment but to its own great surprise, *creates* the dual tartaric crystals? Japp continues, "Only asymmetry can beget asym-

[9] There is known today one clear instance, the reaction of nitrocinnaminacid with bromine where circular-polarized light generates an optically active substance.

metry." The truth of that statement I am willing to admit; but it is of little help since there is no symmetry in the accidental past and present set-up of the actual world which begets the future.

There is however a real difficulty: Why should nature produce only one of the doublets of so many enantiomorphic forms the origin of which most certainly lies in living organisms? Pascual Jordan points to this fact as a support for his opinion that the beginnings of life are not due to chance events which, once a certain stage of evolution is reached, are apt to occur continuously now here now there, but rather to an event of quite singular and improbable character, occurring once by accident and then starting an avalanche by autocatalytic multiplication. Indeed had the asymmetric protein molecules found in plants and animals an independent origin in many places at many times, then their laevo- and dextro-varieties should show nearly the same abundance. Thus it looks as if there is some truth in the story of Adam and Eve, if not for the origin of mankind then for that of the primordial forms of life. It was in reference to these biological facts when I said before that if taken at their face value they suggest an intrinsic difference between left and right, at least as far as the constitution of the organic world is concerned. But we may be sure the answer to our riddle does not lie in any universal biological laws but in the accidents of the genesis of the organismic world. Pascual Jordan shows one way out; one would like to find a less radical one, for instance by reducing the asymmetry of the inhabitants on earth to some inherent, though accidental, asymmetry of the earth itself, or of the light received on earth from the sun.   But neither the earth's rotation nor the combined magnetic fields of earth and sun are of immediate help in this regard. Another possibility would be to assume that development actually started from an equal distribution of the enantiomorph forms, but that this is an unstable equilibrium which under a slight chance disturbance tumbled over.

From the phylogenetic problems of left and right let us finally turn to their onto-genesis.   Two questions arise: Does the first division of the fertilized egg of an animal into two cells fix the median plane, so that one of the cells contains the potencies for its left, the other for its right half?   Secondly what determines the plane of the first division?   I begin with the second question.   The egg of any animal above the protozoa possesses from the beginning a polar axis connecting what develops into the animal and the vegetative poles of the blastula.   This axis together with the point where the fertilizing spermatozoon enters the egg determines a plane, and it would be quite natural to assume that this is the plane of the first division.   And indeed there is evidence that it is so in many cases. Present opinion seems to incline toward the assumption that the primary polarity as well as the subsequent bilateral symmetry come about by external factors actualizing poten-tialities inherent in the genetic constitution. In many instances the direction of the polar axis is obviously determined by the attach-ment of the oozyte to the wall of the ovary, and the point of entrance of the fertilizing sperm is, as we said, at least one, and often the most decisive, of the determining factors for the median plane.   But other agencies may also be responsible for the fixation of the one and the other. In the sea-weed *Fucus* light or electric fields or chemical gradients determine the polar axis, and in some insects and cephalopods the median plane appears to be fixed by ovarian influences before fertilization.[10] The underlying constitution on which these agencies work is sought by some biologists in an intimate preformed structure, of which we do not yet have a clear picture. Thus Conklin has spoken of a spongioplasmic framework, others of a cytoskeleton, and as there is now a strong tendency among biochemists to reduce structural properties to fibers, so much so that Joseph Needham in his Terry Lectures on *Order and life* (1936) dares the aphorism that biology is largely the study of fibers, one may expect them to find that that intimate structure of the egg consists of a framework of elongated protein molecules or fluid crystals.

We know a little more about our first question whether the first mitosis of the cell divides it into left and right. Because of the fundamental character of bilateral symmetry the hypothesis that this is so seems plausible enough. However, the answer cannot be an unqualified affirmation. Even if the hypothesis should be true for the normal development we know from experiments first performed by Hans Driesch on the sea urchin

[10] Julian S. Huxley and G. R. de Beer in their classical *Elements of embryology* (Cambridge University Press, 1934) give this formulation (Chapter XIV, Summary, p. 438): "In the earliest stages, the egg acquires a unitary organization of the gradient-field type in which quantitative differentials of one or more kinds extend across the substance of the egg in one or more directions. The constitution of the egg predetermines it to be able to produce a gradient-field of a particular type; however, the localization of the gradients is not predetermined, but is brought about by agencies external to the egg."

that a single blastomere isolated from its partner in the two-cell stage develops into a whole gastrula differing from the normal one only by its smaller size. Here are Driesch's famous pictures. It must be admitted that this is not so for all species. Driesch's discovery led to the distinction between the actual and the potential destiny of the several parts of an egg. Driesch himself speaks of prospective significance (prospektive Bedeutung), as against prospective potency (prospektive Potenz); the latter is wider than the former, but shrinks in the course of development. Let me illustrate this basic point by another example taken from the determination of limb-buds of amphibia. According to experiments performed by R. G. Harrison, who transplanted discs of the outer wall of

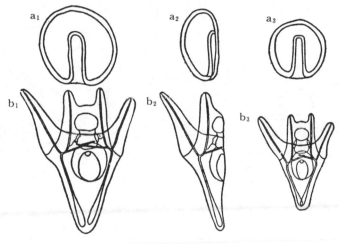

FIG. 17

Experiments on pluripotence in *Echinus*.
a_1 and b_1. Normal gastrula and normal pluteus.
a_2 and b_2. Half-gastrula and half-pluteus, expected by Driesch.
a_3 and b_3. The small but whole gastrula and pluteus, which he actually obtained.

the body representing the buds of future limbs, the antero-posterior axis is determined at a time when transplantation may still invert the dorso-ventral and the medio-lateral axes; thus at this stage the opposites of left and right still belong to the prospective potencies of the discs, and it depends on the influence of the surrounding tissues in which way this potency will be actualized.

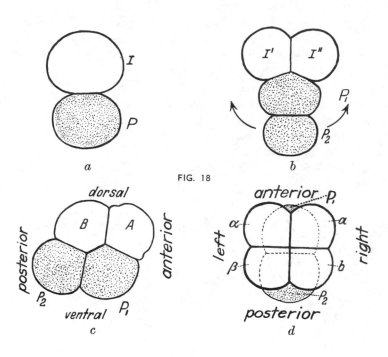

FIG. 18

Driesch's violent encroachment on the normal development proves that the first cell division may not fix left and right of the growing organism for good. But even in normal development the plane of the first division may not be the median. The first stages of cell division have been closely studied for the worm *Ascaris megalocephala*,

36

parts of whose nervous system are asymmetric. First the fertilized egg splits into a cell I and a smaller P of obviously different nature (Fig. 18). In the next stage they divide along two perpendicular planes into $I' + I''$ and $P_1 + P_2$ respectively. Thereafter the handle $P_1 + P_2$ turns about so that P_2 comes into contact with either I' or I''; call the one it contacts B, the other A. We now have a sort of rhomboid and roughly AP_2 is the antero-posterior axis and BP_1 the dorsal-ventral one. Only the next division which along a plane perpendicular to the one separating A and B splits A as well as B into symmetric halves $A = a + \alpha$, $B = b + \beta$, is that which determines left and right. A further slight shift of the configuration destroys this bilateral symmetry. The question arises whether the direction of the two consecutive shifts is a chance event which decides first between anterior and posterior and then between left and right, or whether the constitution of the egg in its one-cell stage contains specific agents which determine the direction of these shifts. The hypothesis of the mosaic egg favoring the second hypothesis seems more likely for the species *Ascaris*.

There are known a number of cases of genotypical inversion where the genetic constitutions of two species are in the same relation as the atomic constitutions of two enantiomorph crystals. More frequent, however, is phenotypical inversion. Left-handedness in man is an example. I give another more interesting one. Several crustacea of the lobster type have two morphologically and functionally different claws, a bigger A and a smaller a. Assume that in normally developed individuals of our species, A is the right claw. If in a young animal you cut off

37

the right claw, inversive regeneration takes place: the left claw develops into the bigger form A while at the place of the right claw a small one of type a is regenerated. One has to infer from such and similar experiences the bipotentiality of plasma, namely that all generative tissues which contain the potency of an asymmetric character have the potency of bringing forth both forms, so however that in normal development always one form develops, the left or the right. Which one is genetically determined, but abnormal external circumstances may cause inversion. On the basis of the strange phenomenon of inversive regeneration Wilhelm Ludwig developed the hypothesis that the decisive factors in asymmetry may not be such specific potencies as, say, the development of a "right claw of type A," but two R and L (right and left) agents which are distributed in the organism with a certain gradient, the concentration of one falling off from right to left, the other in the opposite direction. The essential point is that there is not one but that there are two opposite gradient fields R and L. Which is produced in greater strength is determined by the genetic constitution. If, however, by some damage to the prevalent agent the other previously suppressed one becomes prevalent, then inversion takes place. Being a mathematician and not a biologist I report with the utmost caution on these matters, which seem to me of highly hypothetical nature. But it is clear that the contrast of left and right is connected with the deepest problems concerning the phylogenesis as well as the ontogenesis of organisms.

TRANSLATORY, ROTATIONAL, AND
RELATED SYMMETRIES

TRANSLATORY, ROTATIONAL, AND
RELATED SYMMETRIES

FROM BILATERAL we shall now turn to other kinds of geometric symmetry. Even in discussing the bilateral type I could not help drawing in now and then such other symmetries as the cylindrical or the spherical ones. It seems best to fix the underlying general concept with some precision beforehand, and to that end a little mathematics is needed, for which I ask your patience. I have spoken of transformations. A mapping S of space associates with every space point p a point p' as its image. A special such mapping is the identity I carrying every point p into itself. Given two mappings S, T, one can perform one after the other: if S carries p into p' and T carries p' into p'' then the resulting mapping, which we denote by ST, carries p into p''. A mapping may have an inverse S' such that $SS' = I$ and $S'S = I$; in other words, if S carries the arbitrary point p into p' then S' carries p' back into p, and a similar condition prevails with S' performed in the first and S in the second place. For such a one-to-one mapping S the word transformation was used in the first lecture; let the inverse be denoted by S^{-1}. Of course, the identity I is a transformation, and I itself is its inverse. Reflection in a plane, the basic operation of bilateral symmetry, is such that its iteration SS results in the identity; in other words, it is its own inverse. In general composition of mappings is not commutative; ST need not be the same as TS. Take for instance a point o in a plane and let S be a horizontal translation carrying o into o_1 and T a rotation around o by $90°$. Then ST carries o into the point o_2 (Fig. 19), but TS carries o into o_1. If S is a transformation with the inverse S^{-1}, then S^{-1} is also a transformation and its inverse is S. The composite of two transformations ST is a transformation again, and $(ST)^{-1}$ equals $T^{-1}S^{-1}$ (in this order!). With this rule, although perhaps not with its mathematical expression, you are all familiar. When you dress, it is not immaterial in which order you perform the operations; and when in dressing you start with the shirt and end up with the coat, then in undressing you observe the opposite order; first take off the coat and the shirt comes last.

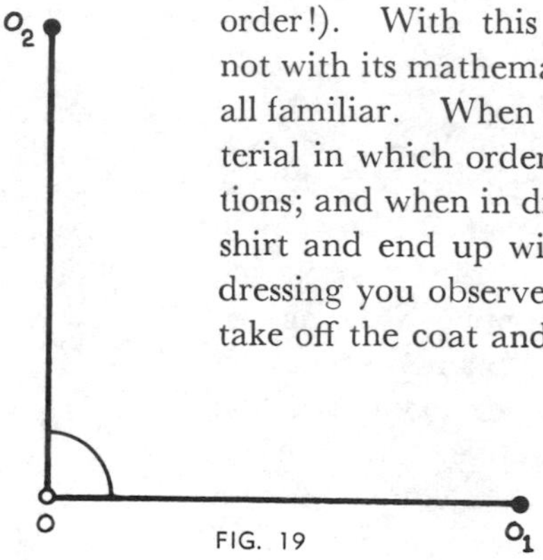

FIG. 19

I have further spoken of a special kind of transformations of space called similarity by the geometers. But I preferred the name of automorphisms for them, defining them with Leibniz as those transformations which leave the structure of space unchanged. For the moment it is immaterial wherein that structure consists. From the very definition it is clear that the identity I is an automorphism, and if S is, so is the inverse S^{-1}. Moreover the composite ST of two automorphisms S, T is again an automorphism. This is only another way of saying that (1) every figure is similar to itself, (2) if figure F' is similar to F then F is similar to F', and (3) if F is similar to F' and F' to F'' then F is similar to F''. The mathematicians have adopted the word *group* to describe this situation and therefore say that the *automorphisms form a group*. Any totality, any set Γ of transformations form a group provided the following conditions are satisfied: (1) the identity I belongs to Γ; (2) if S belongs to Γ then its inverse S^{-1} does; (3) if S and T belong to Γ then the composite ST does.

One way of describing the structure of space, preferred by both Newton and Helmholtz, is through the notion of congruence. Congruent parts of space V, V' are such as can be occupied by the same rigid body in two of its positions. If you move the body from the one into the other position the particle of the body covering a point p of V will afterwards cover a certain point p' of V', and thus the result of the motion is a mapping $p \rightarrow p'$ of V upon V'. We can extend the rigid body either actually or in imagination so as to cover an arbitrarily given point p of space, and hence the congruent mapping $p \rightarrow p'$ can be extended to the entire space. Any such congruent transformation—I call it by that name because it evidently has an inverse $p' \rightarrow p$—is a similarity or an automorphism; you can easily convince yourselves that this follows from the very concepts. It is evident moreover that the congruent transformations form a group, a subgroup of the group of automorphisms. In more detail the situation is this. Among the similarities there are those which do not change the dimensions of a body; we shall now call them congruences. A congruence is either proper, carrying a left screw into a left and a right one into a right, or it is improper or reflexive, changing a left screw into a right one and vice versa. The proper congruences are those transformations which a moment ago we called congruent transformations, connecting the positions of points of a rigid body before and after a motion. We shall now call them simply motions (in a non-kinematic geometric sense) and call the improper congruences reflections, after the most important example: reflection in a plane, by which a body goes over into its mirror image. Thus we have this step-wise arrangement: similarities → congruences = similarities without change of scale → motions = proper congruences. The congruences form a subgroup of the similarities, the motions form a subgroup of the group of congruences, of index 2. The latter addition means that if B is any given improper congruence, we obtain all improper congruences in the form BS by composing B with all possible proper congruences S. Hence the proper congruences form one half, and the improper ones another half, of the group of all congruences. But only the first half is a group; for the composite AB of two improper congruences A, B is a proper congruence.

A congruence leaving the point O fixed may be called *rotation* around O; thus there are proper and improper rotations. The

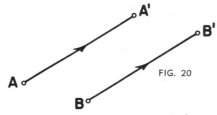

FIG. 20

rotations around a given center O form a group. The simplest type of congruences are the *translations*. A translation may be represented by a vector $\overrightarrow{AA'}$; for if a translation carries a point A into A' and the point B into B' then BB' has the same direction and length as AA', in other words the vector $BB' = AA'$.[1] The translations form a group; indeed the succession of the two translations $\overrightarrow{AB}$, $\overrightarrow{BC}$ results in the translation $\overrightarrow{AC}$.

What has all this to do with symmetry? It provides the adequate mathematical language to define it. Given a spatial configuration $\mathfrak{F}$, those automorphisms of space which leave $\mathfrak{F}$ unchanged form a group Γ,

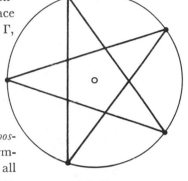

FIG. 21

and *this group describes exactly the symmetry possessed by* $\mathfrak{F}$. Space itself has the full symmetry corresponding to the group of all automorphisms, of all similarities. The symmetry of any figure in space is described by a subgroup of that group. Take for instance the famous pentagram by which Dr. Faust banned Mephistopheles the devil. It is carried into itself by the five proper rotations around its center O, the angles of which are multiples of $360°/5$ (including the identity), and then by the five reflections in the lines joining O with the five vertices. These ten operations form a group, and that group tells us what sort of symmetry the pentagram

[1] While a segment has only length, a vector has length and direction. A vector is really the same thing as a translation, although one uses different phraseologies for vectors and translations. Instead of speaking of the translation $\mathfrak{a}$ which carries the point A into A' one speaks of the vector $\mathfrak{a} = \overrightarrow{AA'}$; and instead of the phrase: the translation $\mathfrak{a}$ carries A into A' one says that A' is the end point of the vector $\mathfrak{a}$ laid off from A. The same vector laid off from B ends in B' if the translation carrying A into A' carries B into B'.

45

possesses. Hence the natural generalization which leads from bilateral symmetry to symmetry in this wider geometric sense consists in replacing reflection in a plane by any group of automorphisms. The circle in a plane with center O and the sphere in space around O have the symmetry described by the group of all plane or spatial rotations respectively.

If a figure $\mathfrak{F}$ does not extend to infinity then an automorphism leaving the figure invariant must be scale-preserving and hence a congruence, unless the figure consists of one point only. Here is the simple proof. Had we an automorphism leaving $\mathfrak{F}$ unchanged, but changing the scale, then either this automorphism or its inverse would increase (and not decrease) all linear dimensions in a certain proportion $a:1$ where a is a number greater than 1. Call that automorphism S, and let α, β be two different points of our figure $\mathfrak{F}$. They have a positive distance d. Iterate the transformation S,

$$S = S^1, \; SS = S^2, \; SSS = S^3, \; \ldots .$$

The n-times iterated transformation S^n carries α and β into two points α_n, β_n of our figure whose distance is $d \cdot a^n$. With increasing exponent n this distance tends to infinity. But if our figure $\mathfrak{F}$ is bounded, there is a number c such that no two points of $\mathfrak{F}$ have a distance greater than c. Hence a contradiction arises as soon as n becomes so large that $d \cdot a^n > c$. The argument shows another thing: Any finite group of automorphisms consists exclusively of congruences. For if it contains an S that enlarges linear dimensions at the ratio $a : 1$, $a > 1$, then all the infinitely many iterations $S^1, S^2, S^3, \cdots$ contained in the group would be different because they enlarge at different scales a^1, a^2,

a^3, $\cdot$ $\cdot$ $\cdot$. For such reasons as these we shall almost exclusively consider groups of congruences—even if we have to do with actually or potentially infinite configurations such as band ornaments and the like.

After these general mathematical considerations let us now take up some special groups of symmetry which are important in art or nature. The operation which defines bilateral symmetry, mirror reflection, is essentially a one-dimensional operation. A straight line can be reflected in any of its points O; this reflection carries a point P into that point P' that has the same distance from O but lies on the other side. Such reflections are the only improper congruences of the one-dimensional line, whereas its only proper congruences are the translations. Reflection in O followed by the translation OA yields reflection in that point A_1 which halves the distance OA. A figure which is invariant under a translation t shows what in the art of ornament is called "infinite rapport," i.e. repetition in a regular spatial rhythm. A pattern invariant under the translation t is also invariant under its iterations t^1, t^2, t^3, $\cdot$ $\cdot$ $\cdot$, moreover under the identity $t^0 = I$, and under the inverse t^{-1} of t and its iterations t^{-1}, t^{-2}, t^{-3}, $\cdot$ $\cdot$ $\cdot$. If t shifts the line by the amount a then t^n shifts it by the amount

$$na \qquad (n = 0, \pm 1, \pm 2, \cdot \cdot \cdot).$$

Hence if we characterize a translation t by the shift a it effects then the iteration or power t^n is characterized by the multiple na. All translations carrying into itself a given pattern of infinite rapport on a straight line are in this sense multiples na of one basic translation a. This rhythmic may be combined with reflexive symmetry. If so the

47

centers of reflections follow each other at half the distance $\frac{1}{2}a$. Only these two types of symmetry, as illustrated by Fig. 22, are possible for a one-dimensional pattern or "ornament." (The crosses $\times$ mark the centers of reflection.)

FIG. 22

Of course the real band ornaments are not strictly one-dimensional, but their symmetry as far as we have described it now makes use of their longitudinal dimension only. Here are some simple examples from Greek art. The first (Fig. 23) which shows a very frequent motif, the palmette, is of type I (translation + reflection). The next (Fig. 24) are without reflections (type II). This frieze of Persian bowmen from Darius' palace in Susa (Fig. 25) is pure translation; but you should notice that the basic translation covers twice the distance from man to man because the costumes of the bowmen alternate. Once more I shall point out the

FIG. 23

FIG. 24

Monreale mosaic of the Lord's Ascension (Fig. 10), but this time drawing your attention to the band ornaments framing it. The widest, carried out in a peculiar technique, later taken up by the Cosmati, displays the translatory symmetry only by repetition of the outer contour of the basic tree-like motif,

FIG. 25

while each copy is filled by a different highly symmetric two-dimensional mosaic. The palace of the doges in Venice (Fig. 26) may stand for translatory symmetry in architecture. Innumerable examples could be added.

As I said before, band ornaments really consist of a two-dimensional strip around a central line and thus have a second trans-

FIG. 26

versal dimension. As such they can have further symmetries. The pattern may be carried into itself by reflection in the central line l; let us distinguish this as longitudinal reflection from the transversal reflection in a line perpendicular to l. Or the pattern may be carried into itself by longitudinal reflection combined with the translation by $\frac{1}{2}a$ (longitudinal slip reflection). A frequent motif in band ornaments are cords, strings, or plaits of some sort, the design of which suggests that one strand crosses the other in space (and thus makes part of it invisible). If this interpretation is accepted, further operations become possible; for example, reflection in the plane of the ornament would change a strand slightly above the plane into one below. All this can be thoroughly analyzed in terms of group theory as is for instance done in a section of Andreas Speiser's book, *Theorie der Gruppen von endlicher Ordnung*, quoted in the Preface.

In the organic world the translatory symmetry, which the zoologists call metamerism, is seldom as regular as bilateral symmetry frequently is. A maple shoot and a shoot of *Angraecum distichum* (Fig. 27) may serve as examples.[2] In the latter case translation is accompanied by longitudinal slip reflection. Of course the pattern does not go on into infinity (nor does a band ornament), but one may say that it is potentially infinite at least in one direction, as in the course of time ever new segments separated from each other by a bud come into being. Goethe said of the tails of vertebrates that they allude as it were to the potential infinity of organic existence. The central part of the animal shown in this picture, a scolopendrid (Fig. 28), possesses fairly regular translational, combined with bilateral, symmetry, the basic operations of which are translation by one segment and longitudinal reflection.

In one-dimensional *time* repetition at equal intervals is the musical principle of *rhythm*. As a shoot grows it translates, one might say, a slow temporal into a spatial rhythm. Re-

FIG. 27

FIG. 28

[2] This and the next picture are taken from *Studium Generale*, p. 249 and p. 241 (article by W. Troll, "Symmetriebetrachtung in der Biologie").

flection, inversion in time, plays a far less important part in music than rhythm does. A melody changes its character to a considerable degree if played backward, and I, who am a poor musician, find it hard to recognize reflection when it is used in the construction of a fugue; it certainly has no such spontaneous effect as rhythm. All musicians agree that underlying the emotional element of music is a strong formal element. It may be that it is capable of some such mathematical treatment as has proved successful for the art of ornaments. If so, we have probably not yet discovered the appropriate mathematical tools. This would not be so surprising. For after all, the Egyptians excelled in the ornamental art four thousand years before the mathematicians discovered in the group concept the proper mathematical instrument for the treatment of ornaments and for the derivation of their possible symmetry classes. Andreas Speiser, who has taken a special interest in the group-theoretic aspect of ornaments, tried to apply combinatorial principles of a mathematical nature also to the formal problems of music. There is a chapter with this title in his book, "Die mathematische Denkweise," (Zurich, 1932). As an example, he analyzes Beethoven's pastoral sonata for piano, opus 28, and he also points to Alfred Lorenz's investigations on the formal structure of Richard Wagner's chief works. Metrics in poetry is closely related, and here, so Speiser maintains, science has penetrated much deeper. A common principle in music and prosody seems to be the configuration *a a b* which is often called a bar: a theme *a* that is repeated and then followed by the "envoy" *b*; strophe, antistrophe, and epode in Greek choric lyrics.

But such schemes fall hardly under the heading of symmetry.[3]

We return to symmetry in space. Take a band ornament where the individual section repeated again and again is of length a and sling it around a circular cylinder, the circumference of which is an integral multiple of a, for instance $25a$. You then obtain a pattern which is carried over into itself through the rotation around the cylinder axis by $\alpha = 360°/25$ and its repetitions. The twenty-fifth iteration is the rotation by $360°$, or the identity. We thus get a finite group of rotations of order 25, i.e. one consisting of 25 operations. The cylinder may be replaced by any surface of cylindrical sym-

[3] The reader should compare what G. D. Birkhoff has to say on the mathematics of poetry and music in the two publications quoted in Lecture I, note 1.

FIG. 29   FIG. 30

53

metry, namely by one that is carried into itself by all rotations around a certain axis, for instance by a vase. Fig. 29 shows an attic vase of the geometric period which displays quite a number of simple ornaments of this type. The principle of symmetry is the same, although the style is no longer "geometric," in this Rhodian pitcher (Fig. 30), Ionian school of the seventh century B.C. Other illustrations are such capitals as these from early Egypt (Fig. 31). Any finite group of proper rotations around a point O in a plane, or around a given axis in space, contains a primitive rotation t whose angle is an aliquot part $360°/n$ of the full rotation by $360°$, and consists of its iterations t^1, t^2, . . . , t^{n-1}, $t^n =$ identity. The order n completely characterizes this group. The result follows from the analogous fact that any group of translations of a line, provided it contains no operations arbitrarily near to the identity except the identity itself, consists of the itera-

FIG. 31

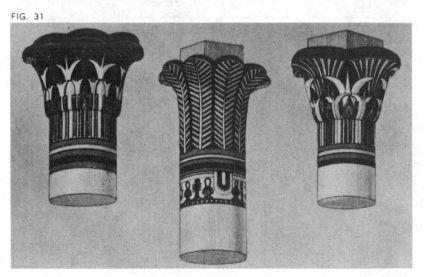

54

FIG. 32

tions νa of a single translation a ($\nu = 0$, ± 1, ± 2, $\cdots$).

The wooden dome in the Bardo of Tunis, once the palace of the Beys of Tunis (Fig. 32), may serve as an example from interior architecture. The next picture (Fig. 33) takes you to Pisa; the Baptisterium with the tiny-looking statue of John the Baptist on top is a central building in whose exterior you can distinguish six horizontal layers each of rotary symmetry of a different order n. One could make the picture still more impressive by adding the leaning tower with its six galleries of arcades all having rotary symmetry

FIG. 33

of the same high order and the dome itself, the exterior of whose nave displays in columns and friezes patterns of the lineal translatory type of symmetry while the cupola is surrounded by a colonnade of high order rotary symmetry.

An entirely different spirit speaks to us from the view, seen from the rear of the choir, of the Romanesque cathedral in Mainz, Germany (Fig. 34). Yet again repetition in the round arcs of the friezes, octagonal central symmetry ($n = 8$, a low value compared to those embodied in the several layers of the Pisa Baptisterium) in the small rosette and the three towers, while bilateral symmetry rules the structure as a whole as well as almost every detail.

Cyclic symmetry appears in its simplest form if the surface of fully cylindrical symmetry is a plane perpendicular to the axis. We then can limit ourselves to the two-

FIG. 34

FIG. 35

dimensional plane with a center O. Magnificent examples of such central plane symmetry are provided by the rose windows of Gothic cathedrals with their brilliant-colored glasswork. The richest I remember is the rosette of St. Pierre in Troyes, France, which is based on the number 3 throughout.

Flowers, nature's gentlest children, are also conspicuous for their colors and their cyclic symmetry. Here (Fig. 35) is a picture of an iris with its triple pole. The symmetry of 5 is most frequent among flowers. A page like the following (Fig. 36) from Ernst

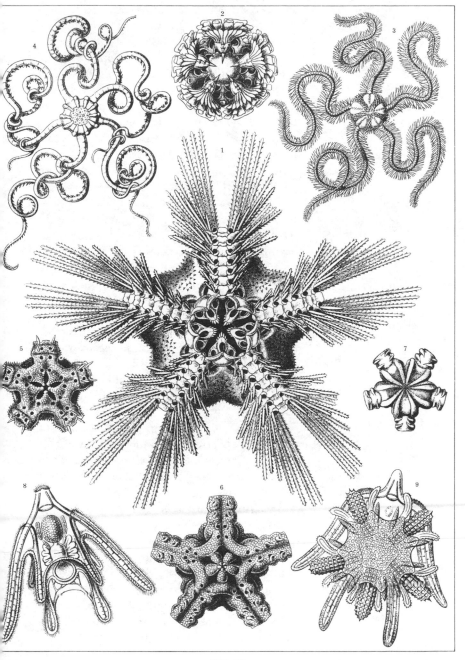

FIG. 36

Haeckel's *Kunstformen der Natur* seems to indicate that it also occurs not infrequently among the lower animals. But the biologists warn me that the outward appearance of these echinoderms of the class of *Ophiodea* is to a certain degree deceptive; their larvae are organized according to the principle of bilateral symmetry. No such objection attaches to the next picture from the same source (Fig. 37), a *Discomedusa* of octagonal symmetry. For the coelentera occupy a place in the phylogenetic evolution where cyclic has not yet given way to bilateral symmetry. Haeckel's extraordinary work, in which his interest in the concrete forms of organisms finds expression in countless drawings executed in minutest detail, is a true nature's codex of symmetry. Equally revealing for Haeckel, the biologist, are the thousands and thousands of figures in his *Challenger Monograph*, in which he describes for the first time 3,508 new species of radiolarians discovered by him on the Challenger Expedition, 1887. One should not forget these accomplishments over the often all-too-speculative phylogenetic constructions in which this enthusiastic apostle of Darwinism indulged, and over his rather shallow materialistic philosophy of monism, which made quite a splash in Germany around the turn of the century.

Speaking of *Medusae* I cannot resist the temptation of quoting a few lines from D'Arcy Thompson's classic work on *Growth and Form*, a masterpiece of English literature, which combines profound knowledge in geometry, physics, and biology with humanistic erudition and scientific insight of unusual originality. Thompson reports on physical experiments with hanging drops which serve

FIG. 37

FIG. 38

to illustrate by analogy the formation of medusae. "The living medusa," he says, "has geometrical symmetry so marked and regular as to suggest a physical or mechanical element in the little creatures' growth and construction. It has, to begin with, its vortex-like bell or umbrella, with its symmetrical handle or manubrium. The bell is traversed by radial canals, four or in multiples of four; its edge is beset with tentacles, smooth or often beaded, at regular intervals or of graded sizes; and certain sensory structures, including solid concretions or 'otoliths,' are also symmetrically interspersed. No sooner made, then it begins to pulsate; the bell begins to 'ring.' Buds, miniature replicas of the parent-organism, are very apt to appear on the tentacles, or on the manubrium or sometimes on the edge of the bell; we seem to see one vortex producing others before our eyes. The development of a medusoid deserves to be studied without prejudice from this point of view. Certain it is that the tiny medusoids of *Obelia*, for instance, are budded off with a rapidity and a complete perfection which suggests an automatic and all but instantaneous act of conformation, rather than a gradual process of growth."

While pentagonal symmetry is frequent in the organic world, one does not find it among the most perfectly symmetrical creations of inorganic nature, among the crystals. There no other rotational symmetries are possible than those of order 2, 3, 4, and 6. Snow crystals provide the best known specimens of hexagonal symmetry. Fig. 38 shows some of these little marvels of frozen water. In my youth, when they came down from heaven around Christmastime blanketing the landscape, they were the delight of old and young. Now only the skiers like them, while they have become the abomination of motorists. Those versed in English literature will remember Sir Thomas Browne's quaint account in his *Garden of Cyrus* (1658) of hexagonal and "quincuncial" symmetry which "doth neatly declare how nature Geometrizeth and observeth order in all things." One versed in German literature will remember how Thomas Mann in his *Magic Mountain*[4] describes the "hexagonale Unwesen" of the snow storm in which his hero, Hans Castorp, nearly perishes when he falls asleep with exhaustion and leaning against a barn dreams his deep dream of death and love. An hour before when Hans sets out on his unwarranted expedition on skis he enjoys the play of the flakes "and among these myriads of enchanting little stars," so he philosophizes, "in their hidden splendor, too small for man's naked eye to see, there was not one like unto another; an endless inventiveness governed the development and unthinkable differentiation of one and the same basic scheme, the equilateral, equiangled hexagon. Yet each in itself—this was the uncanny, the antiorganic, the life-denying character of them all—each of them was absolutely symmetrical, icily regular in form. They were too regular, as substance adapted to life never was to this degree—the living principle shuddered at this perfect precision, found it deathly, the very marrow of death— Hans Castorp felt he understood now the reason why the builders of antiquity purposely and secretly introduced minute varia-

[4] I quote Helen Lowe-Porter's translation, Knopf, New York, 1927 and 1939.

tion from absolute symmetry in their columnar structures."[5]

Up to now we have paid attention to proper rotations only. If improper rotations are taken into consideration, we have the two following possibilities for finite groups of rotations around a center O in plane geometry, which correspond to the two possibilities we encountered for ornamental symmetry on a line: (1) the group consisting of the repetitions of a single proper rotation by an aliquot part $\alpha = 360°/n$ of $360°$; (2) the group of these rotations combined with the reflections in n axes forming angles of $\frac{1}{2}\alpha$. The first group is called the cyclic group C_n and the second the dihedral group D_n. Thus these are the only possible central symmetries in two-dimensions:

(1) $C_1, C_2, C_3, \cdots ; D_1, D_2, D_3, \cdots .$

C_1 means no symmetry at all, D_1 bilateral symmetry and nothing else. In architecture the symmetry of 4 prevails. Towers often have hexagonal symmetry. Central buildings with the symmetry of 6 are much less frequent. The first pure central building after antiquity, S. Maria degli Angeli in Florence (begun 1434), is an octagon. Pentagons are very rare. When once before I lectured on symmetry in Vienna in 1937 I said I knew of only one example and that a very inconspicuous one, forming the passageway from San Michele di Murano in Venice

[5] Dürer considered his canon of the human figure more as a standard from which to deviate than as a standard toward which to strive. Vitruvius' *temperaturae* seem to have the same sense, and maybe the little word "almost" in the statement ascribed to Polykleitos and mentioned in Lecture I, note 1, points in the same direction.

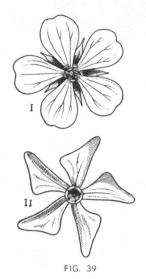

I

II

FIG. 39

FIG. 40

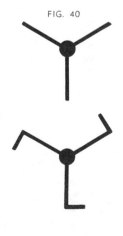

to the hexagonal Capella Emiliana. Now, of course, we have the Pentagon building in Washington. By its size and distinctive shape, it provides an attractive landmark for bombers. Leonardo da Vinci engaged in systematically determining the possible symmetries of a central building and how to attach chapels and niches without destroying the symmetry of the nucleus. In abstract modern terminology, his result is essentially our above table of the possible finite groups of rotations (proper and improper) in two dimensions.

So far the rotational symmetry in a plane had always been accompanied by reflective symmetry; I have shown you quite a number of examples for the dihedral group D_n and none for the simpler cyclic group C_n. But this is more or less accidental. Here (Fig. 39) are two flowers, a *geranium* (I) with the symmetry group D_5, while *Vinca herbacea* (II) has the more restricted group C_5 owing to the asymmetry of its petals. Fig. 40 shows what is perhaps the simplest figure with rotational symmetry, the tripod ($n = 3$). When one wants to eliminate the attending reflective symmetry, one puts little flags unto the arms and obtains the triquetrum, an old magic symbol. The Greeks, for instance, used it with the Medusa's head in the center as the symbol for the three-cornered Sicily. (Mathematicians are familiar with it as the seal on the cover of the *Rendiconti del Circolo Matematico di Palermo*.) The modification with four instead of three arms is the swastika, which need not be shown here—one of the most primeval symbols of mankind, common possession of a number of apparently independent civilizations. In my lecture on symmetry in Vienna in the fall of 1937, a

FIG 41

short time before Hitler's hordes occupied Austria, I added concerning the swastika: "In our days it has become the symbol of a terror far more terrible than the snake-girdled Medusa's head"—and a pandemonium of applause and booing broke loose in the audience. It seems that the origin of the magic power ascribed to these patterns lies in their startling incomplete symmetry— rotations without reflections. Here (Fig. 41) is the gracefully designed staircase of the pulpit of the Stephan's dome in Vienna; a triquetrum alternates with a swastika-like wheel.

So much about rotational symmetry in two dimensions. If dealing with potentially infinite patterns like band ornaments or with infinite groups, the operation under which the pattern is invariant is not of necessity a congruence but could be a similarity. A similarity in one dimension that is not a mere translation has a fixed point O and is a dilatation s from O in a certain ratio $a:1$ where $a \neq 1$. It is no essential restriction to assume $a > 0$. Indefinite iteration of this operation generates a group Σ consisting of the dilatations

$$(2) \quad s^n \qquad (n = 0, \pm 1, \pm 2, \cdots).$$

FIG. 42

A good example of this type of symmetry is shown by the shell of *Turritella duplicata* (Fig. 42). It is really quite remarkable how exactly the widths of the consecutive whorls of this shell follow the law of geometric progression.

The hands of some clocks perform a continuous uniform rotation, others jump from minute to minute. The rotations by an integral number of minutes form a discontinuous subgroup within the continuous group of all rotations, and it is natural to consider a rotation s and its iterations (2) as contained in the continuous group. We can apply this viewpoint to any similarity in 1, 2, or 3 dimensions, as a matter of fact to any transformation s. The continuous motion of a space-filling substance, a "fluid," can mathematically be described by giving the transformation $U(t,t')$ which carries the position P_t of any point of the fluid at the moment t over into its position $P_{t'}$ at the time t'. These transformations form a one-parameter group if $U(t,t')$ depends on the time difference $t' - t$ only, $U(t,t') = S(t' - t)$,

68

i.e. if during equal time intervals always the same motion is repeated. Then the fluid is in "uniform motion." The simple group law

$$S(t_1)S(t_2) = S(t_1 + t_2)$$

expresses that the motions during two consecutive time intervals t_1, t_2 result in the motion during the time $t_1 + t_2$. The motion during 1 minute leads to a definite transformation $s = S(1)$, and for all integers n the motion $S(n)$ performed during n minutes is the iteration s^n: the discontinuous group Σ consisting of the iterations of s is embedded in the continuous group with the parameter t consisting of the motions $S(t)$. One could say that the continuous motion consists of the endless repetition of the same infinitesimal motion in consecutive infinitely small time intervals of equal length.

We could have applied this consideration to the rotations of a plane disc as well as to dilatations. We now envisage any proper similarity s, i.e. one which does not interchange left and right. If, as we assume, it is not a mere translation, it has a fixed point O and consists of a rotation about O combined with a dilatation from the center O. It can be obtained as the stage $S(1)$ reached after 1 minute by a continuous process $S(t)$ of combined uniform rotation and expansion. This process carries a point $\neq O$ along a so-called logarithmic or equiangular spiral. This curve, therefore, shares with straight line and circle the important property of going over into itself by a continuous group of similarities. The words by which James Bernoulli had the *spira mirabilis* adorned on his tombstone in the Münster at Basle, "Eadem mutata resurgo," are a grandilo-

FIG. 43

quent expression of this property. Straight line and circle are limiting cases of the logarithmic spiral, which arise when in the combination rotation-plus-dilatation one of the two components happens to be the identity. The stages reached by the process at the times

$$(3) \quad t = n = \cdots, -2, -1, 0, 1, 2, \cdots$$

form the group consisting of the iterations (2). The well-known shell of *Nautilus* (Fig. 43) shows this sort of symmetry to an astonishing perfection. You see here not only the continuous logarithmic spiral, but the potentially infinite sequence of chambers has a symmetry described by the discontinuous group Σ. For everybody looking at this picture (Fig. 44) of a giant sunflower, *Helianthus maximus*, the florets will naturally arrange themselves into logarithmic spirals, two sets of spirals of opposite sense of coiling.

The most general rigid motion in three-dimensional space is a screw motion s, combination of a rotation around an axis with a translation along that axis.   Under the influence of the corresponding continuous uniform motion any point not on the axis describes a screw-line or helix which, of course, could say of itself with the same right as the logarithmic spiral: *eadem resurgo*.   The stages P_n which the moving point reaches at the equi-

FIG. 44

distant moments (3) are equidistributed over the helix like stairs on a winding staircase. If the angle of rotation of the operation s is a fraction μ/ν of the full angle 360° expressible in terms of small integers μ, ν then every νth point of the sequence P_n lies on the same vertical, and μ full turnings of the screw are necessary to get from P_n to the point $P_{n+\nu}$ above it. The leaves around the shoot of a plant often show such a regular spiral arrangement. Goethe spoke of a spiral tendency in nature, and under the name of *phyllotaxis* this phenomenon, since the days of Charles Bonnet (1754), has been the subject of much investigation and more speculation among botanists.[6] One has found that the fractions μ/ν representing the screw-like arrangement of leaves quite often are members of the "Fibonacci sequence"

(4) $\frac{1}{1}, \frac{1}{2}, \frac{2}{3}, \frac{3}{5}, \frac{5}{8}, \frac{8}{13}, \frac{13}{21}, \frac{21}{34}, \cdots,$

which results from the expansion into a continued fraction of the irrational number $\frac{1}{2}(\sqrt{5} - 1)$. This number is no other but the ratio known as the *aurea sectio*, which has played such a role in attempts to reduce beauty of proportion to a mathematical formula. The cylinder on which the screw is wound could be replaced by a cone; this amounts to replacing the screw motion s by any proper similarity—rotation combined with dilatation. The arrangement of scales on a fir-cone falls under this slightly more general form of symmetry in phyllotaxis. The transition from cylinder over cone to disc is obvious, illustrated by the cylindrical

[6] This phenomenon plays also a role in J. Hambidge's constructions. His *Dynamic symmetry* contains on pp. 146–157 detailed notes by the mathematician R. C. Archibald on the logarithmic spiral, golden section, and the Fibonacci series.

stem of a plant with its leaves, a fir-cone with its scales, and the discoidal inflorescence of *Helianthus* with its florets. Where one can check the numbers (4) best, namely for the arrangement of scales on a fir-cone, the accuracy is not too good nor are considerable deviations too rare. P. G. Tait, in the *Proceedings of the Royal Society of Edinburgh* (1872), has tried to give a simple explanation, while A. H. Church in his voluminuous treatise *Relations of phyllotaxis to mechanical laws* (Oxford, 1901–1903) sees in the arithmetics of phyllotaxis an organic mystery. I am afraid modern botanists take this whole doctrine of phyllotaxis less seriously than their forefathers.

Apart from reflection all symmetries so far considered are described by a group consisting of the iterations of one operation s. In one case, and that is undoubtedly the most important, the resulting group is finite, namely if one takes for s a rotation by an angle $\alpha = 360°/n$ which is an aliquot part of the full rotation $360°$. For the two-dimensional plane there are no other finite groups of proper rotations than these; witness the first line, C_1, C_2, C_3, $\cdots$ of Leonardo's table (1). The simplest figures which have the corresponding symmetry are the regular polygons: the regular triangle, the square, the regular pentagon, etc. The fact that there is for every number $n = 3, 4, 5, \cdots$ a regular polygon of n sides is closely related to the existence for every n of a rotational group of order n in plane geometry. Both facts are far from trivial. Indeed, the situation in three dimensions is altogether different: there do not exist infinitely many regular polyhedra in 3-space, but not more than five, often called the Platonic solids because they play an eminent role in Plato's natural philosophy. They are the regular tetrahedron, the cube, the octahedron, moreover the pentagondodecahedron, the sides of which are twelve regular pentagons, and the icosahedron bounded by twenty regular triangles. One might say that the existence of the first three is a fairly trivial geometric fact. But the discovery of the last two is certainly one of the most beautiful and singular discoveries made in the whole history of mathematics. With a fair amount of certainty, it can be traced to the colonial Greeks in southern Italy. The suggestion has been made that they abstracted the regular dodecahedron from the crystals of pyrite, a sulphurous mineral abundant in Sicily. But as mentioned before, the symmetry of 5 so characteristic for the regular dodecahedron contradicts the laws of crystallography, and indeed one finds that the pentagons bounding the dodecahedra in which pyrite crystallizes have 4 edges of equal, but one of different, length. The first exact construction of the regular pentagondodecahedron is probably due to Theaetetus. There is some evidence that dodecahedra were used as dice in Italy at a very early time and had some religious significance in Etruscan culture. Plato, in the dialogue *Timaeus*, associates the regular pyramid, octahedron, cube, icosahedron, with the four elements of fire, air, earth, and water (in this order), while in the pentagondodecahedron he sees in some sense the image of the universe as a whole. A. Speiser has advocated the view that the construction of the five regular solids is the chief goal of the deductive system of geometry as erected by the Greeks and canonized in Euclid's *Elements*. May I mention, however, that the

Greeks never used the word "symmetric" in our modern sense. In common usage σύμμετρος means *proportionate*, while in Euclid it is equivalent to our *commensurable:* side and diagonal of a square are incommensurable quantities, ἀσύμμετρα μεγέθη.

Here (Fig. 45) is a page from Haeckel's *Challenger monograph* showing the skeletons

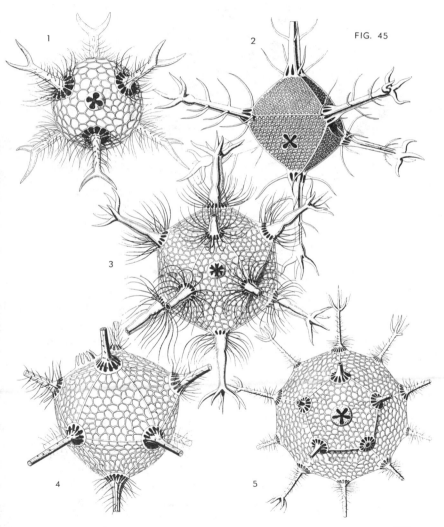

FIG. 45

of several Radiolarians. Nr. 2, 3, and 5 are octahedron, icosahedron, and dodecahedron in astonishingly regular form; 4 seems to have a lower symmetry.

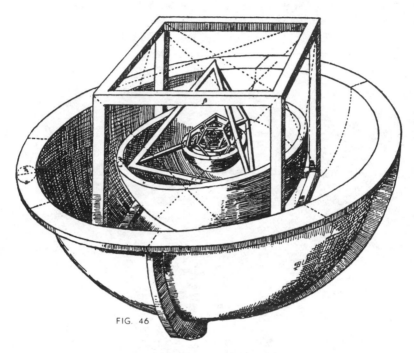

FIG. 46

Kepler, in his *Mysterium cosmographicum*, published in 1595, long before he discovered the three laws bearing his name today, made an attempt to reduce the distances in the planetary system to regular bodies which are alternatingly inscribed and circumscribed to spheres. Here (Fig. 46) is his construction, by which he believed he had penetrated deeply into the secrets of the Creator. The six spheres correspond to the six planets, Saturn, Jupiter, Mars, Earth, Venus, Mercurius, separated in this order by cube, tetrahedron, dodecahedron, octahedron, icosahedron. (Of course, Kepler did not know about the three outer planets, Uranus, Neptune, and Pluto, which were discovered in 1781, 1846, and 1930 respectively.) He tries to find the reasons why the Creator had chosen this order of the Platonic solids and draws parallels between the properties of the planets (astrological rather than astrophysical properties) and those of the corresponding regular bodies. A mighty hymn in which he proclaims his credo, "Credo spatioso numen in orbe," concludes his book. We still share his belief in a mathematical harmony of the universe. It has withstood the test of ever widening experience. But we no longer seek this harmony in static forms like the regular solids, but in dynamic laws.

As the regular polygons are connected with the finite groups of plane rotations, so must the regular polyhedra be intimately related to the finite groups of proper rotations around a center O in space. From the study of plane rotations we at once obtain two types of proper rotation groups in space. Indeed, the group C_n of proper rotations in a horizontal plane around a center O can be interpreted as consisting of rotations in space around the vertical axis through O. Reflection of the horizontal plane in a line l of the plane can be brought about in space through a rotation around l by 180° (*Umklappung*). You may remember that we mentioned this in connection with the analysis of a Sumerian picture (Fig. 4). In this way the group D_n in the horizontal plane is changed into a group D'_n of proper rotations in space; it contains the rotations around a vertical axis through O by the multiples of 360°/n and the Umklappungen around n horizontal axes through O which form equal angles of

$360°/2n$ with each other. But it should be observed that the group D'_1 as well as C_2 consists of the identity and the Umklappung around one line. These two groups are therefore identical, and in a complete list of the *different* groups of proper rotations in three dimensions, D'_1 should be omitted if C_2 is kept. Hence we start our list thus:

$$C_1, C_2, C_3, C_4, \cdots ;$$
$$D'_2, D'_3, D'_4, \cdots .$$

D'_2 is the so-called four-group consisting of the identity and the Umklappungen around three mutually perpendicular axes.

For each one of the five regular bodies we can construct the group of those proper rotations which carry that body into itself. Does this give rise to five new groups? No, only to three, and that for the following reason. Inscribe a sphere into a cube and an octahedron into the sphere such that the corners of the octahedron lie where the sides of the cube touch the sphere, namely in the centers of the six square sides. (Fig. 47 shows the two-dimensional analogue.) In this position cube and octahedron are polar figures in the sense of projective geometry. It is clear that every rotation which carries the cube into itself also leaves the octahedron invariant, and vice versa. Hence the group for the octahedron is the same as for the cube. In the same manner pentagondodecahedron and icosahedron are polar figures. The figure polar to a regular tetrahedron is a regular tetrahedron the corners of which are the antipodes of those of the first. Thus we find three new groups of proper rotations, T, W, and P; they are those leaving invariant the regular tetrahedron, the cube (or octahedron), and the pentagondodecahedron (or

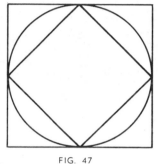

FIG. 47

78

icosahedron) respectively. Their orders, i.e. the number of operations in each of them, are 12, 24, 60 respectively.

It can be shown by a relatively simple analysis (Appendix A) that with the addition of these three groups our table is complete:

$$
\begin{aligned}
&C_n \quad (n = 1, 2, 3, \cdots), \\
(5) \quad &D'_n \quad (n = 2, 3, \cdots); \\
&T, W, P.
\end{aligned}
$$

This is the modern equivalent to the tabulation of the regular polyhedra by the Greeks. These groups, in particular the last three, are an immensely attractive subject for geometric investigation.

What further possibilities arise if improper rotations are also admitted to our groups? This question is best answered by making use of one quite singular improper rotation, namely reflection in O; it carries any point P into its antipode P' with respect to O found by joining P with O and prolonging the straight line PO by its own length: $PO = OP'$. This operation Z commutes with every rotation S, $ZS = SZ$. Now let Γ be one of our finite groups of proper rotations. One way of including improper rotations is simply by adjoining Z, more precisely by adding to the proper rotations S of Γ all the improper rotations of the form ZS (with S in Γ). The order of the group $\bar{\Gamma} = \Gamma + Z\Gamma$ thus obtained is clearly twice that of Γ. Another way of including improper rotations arises from this situation: Suppose Γ is contained as a subgroup of index 2 in another group Γ' of proper rotations; so that one-half of the elements of Γ' lie in Γ, call them S, and one-half, S', do not. Now replace these latter by the improper rotations ZS'. In this manner you get a group $\Gamma'\Gamma$ which contains

Γ while the other half of its operations are improper. For instance, $\Gamma = C_n$ is a subgroup of index 2 of $\Gamma' = D'_n$; the operations S' of D'_n not contained in C_n are the Umklappungen around the n horizontal axes. The corresponding ZS' are the reflections in the vertical planes perpendicular to these axes. Thus $D'_n C_n$ consists of the rotations around the vertical axis, the angles of which are multiples of $360°/n$, and of the reflections in vertical planes through this axis forming angles of $360°/2n$ with each other. You might say that this is the group formerly denoted by D_n. Another example, the simplest of all: $\Gamma = C_1$ is contained in $\Gamma' = C_2$. The one operation S' of C_2 not contained in C_1 is the rotation by $180°$ about the vertical axis; ZS' is reflection in the horizontal plane through O. Hence $C_2 C_1$ is the group consisting of the identity and of the reflection in a given plane; in other words, the group to which bilateral symmetry refers.

The two ways described are the only ones by which improper rotations may be included in our groups. (For the proof see Appendix B.) Hence this is the complete table of all finite groups of (proper and improper) rotations:

$C_n, \quad \bar{C}_n, \quad C_{2n}C_n \quad (n = 1, 2, 3, \cdots)$

$D'_n, \quad \bar{D}'_n, \quad D'_n C_n, \quad D'_{2n}D'_n$
$\qquad\qquad\qquad\qquad\qquad (n = 2, 3, \cdots)$

$T, \quad W, \quad P; \quad \bar{T}, \quad \bar{W}, \quad \bar{P}; \quad WT.$

The last group WT is made possible by the fact that the tetrahedral group T is a subgroup of index 2 of the octahedral group W.

This list will be of importance to us when in the last lecture we shall consider the symmetry of crystals.

ORNAMENTAL SYMMETRY

ORNAMENTAL SYMMETRY

THIS LECTURE will have a more systematic character than the preceding one, in as much as it will be dedicated to one special kind of geometric symmetry, the most complicated but also the most interesting from every angle. In two dimensions the art of surface ornaments deals with it, in three dimensions it characterizes the arrangement of atoms in a crystal. We shall therefore call it the ornamental or crystallographic symmetry.

Let us begin with an ornamental pattern in two dimensions which probably occurs more frequently than any other, both in art and nature: the hexagonal pattern so often used for tiled floors in bathrooms. You see it here realized by the honeycomb as it is built by our common hivebees (Fig. 48). The bees' cells have prismatic shape, the photograph is taken in the direction of these prisms. As a matter of fact, a honeycomb consists of two layers of such cells, the prisms of the one layer facing one way, the other the opposite way. How the inner ends of these two layers dovetail is a spatial problem which we shall presently take up. At the moment we are concerned with the simpler two-dimensional question. If you pile round-shots or round beads in a heap they will of themselves get arranged in the three-dimensional analogue of the hexagonal configuration. In two dimensions the task is to pack equal circles as compactly as possible. You start with a horizontal row of circles that touch each other. If you drop another circle from above upon this row it will nest between two adjacent circles of the row, and the centers of the three circles will form an equilateral triangle. From this upper circle there derives a second horizontal row of circles nesting between those of the first row; and so on (Fig. 49). The circles leave little lacunae between them. The tangents of a circle at the points where it touches the six surrounding circles form a regular hexagon

FIG. 48

circumscribing the circle, and if you replace each circle by this hexagon you obtain the regular configuration of hexagons filling the whole plane.

According to the laws of capillarity a soap film spanned into a given contour made of thin wire assumes the shape of a minimal surface, i.e. it has smaller area than any other

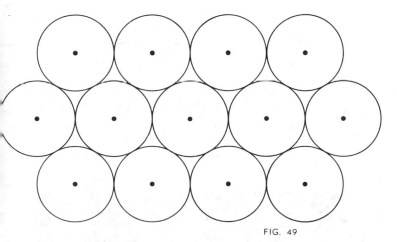

FIG. 49

surface with the same contour. A soap bubble into which a quantum of air is blown assumes spherical form because the sphere encompasses the given volume with a minimum of surface. Thus is it not astonishing that a froth of two-dimensional bubbles of equal area will arrange itself in the hexagonal pattern because among all divisions of the plane into parts of equal area that is the one for which the net of contours has minimum length. We here suppose that the problem has been reduced to two dimensions by dealing with a horizontal layer of bubbles, say, between two horizontal glass plates. If the froth of vesicles has a boundary (an epidermal layer, as the biologist would say), we observe

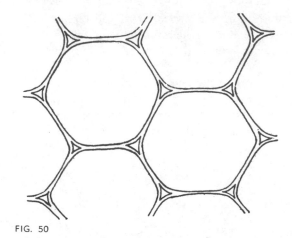

FIG. 50

that it consists of circular arcs each forming
an angle of 120° with the adjacent cell wall
and the next arc, as is required by the law
of minimal length. After this explanation

FIG. 51

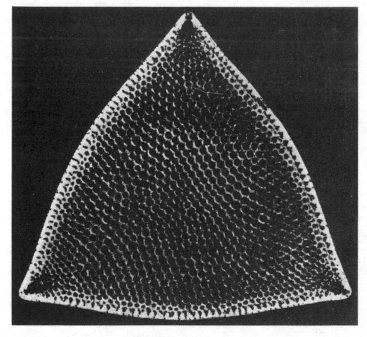

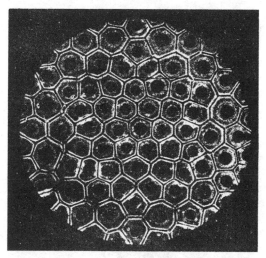

FIG. 52

one will not be surprised to find the hexagonal pattern realized in such different structures as for instance the parenchyma of maize (Fig. 50), the retinal pigment of our eyes, the surface of many diatoms, of which I show here (Fig. 51) a beautiful specimen, and finally the honeycomb. As the bees, which are all of nearly equal size, build their cells gyrating around in them, the cells will form a densest packing of parallel circular cylinders which in cross section appear as our hexagonal pattern of circles. As long as the bees are at work the wax is in a semi-fluid state, and thus the forces of capillarity probably more than the pressures exerted from within by the bees' bodies transform the circles into circumscribed hexagons (whose corners however still show some remains of the circular form). With the parenchym of maize you may compare this artificial cellular tissue (Fig. 52) formed by the diffusion in gelatin of drops of a solution of potassium

FIG. 53

FIG. 54

ferrocyanide. The regularity leaves something to be desired; there are even places where a pentagon is smuggled in instead of a hexagon. Here (Fig. 53 and 54) are two other artificial tissues of hexagonal pattern taken at random from a recent issue of *Vogue* (February 1951). The siliceous skeleton of

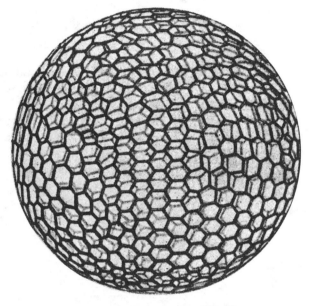

FIG. 55

one of Haeckel's Radiolarians which he called *Aulonia hexagona* (Fig. 55) seems to exhibit a fairly regular hexagonal pattern spread out not in a plane but over a sphere. But a hexagonal net covering the sphere is impossible owing to a fundamental formula of topology. This formula refers to an arbitrary partition of the sphere into countries that border on each other along certain edges. It tells that the number A of countries, the number E of edges and the number C of corners (where at least three countries come together) satisfy the relation $A + C - E = 2$. Now for a hexagonal net we would have $E = 3A$, $C = 2A$ and hence $A + C - E$ $= 0$! And sure enough, we see that some of the meshes of the net of *Aulonia* are not hexagons but pentagons.

From the densest packing of circles in a plane let us now pass to the densest packing of equal spheres, of equal balls in space. We start with one ball and a plane, the "horizontal plane" through its center. In densest packing this ball will touch on twelve others ("like the seeds in a pomegranate," as Kepler says), six in the horizontal plane, three below, and three above.[1] If mutual penetration is impossible the balls in this arrangement, under uniform expansion around their fixed centers, will change into rhombic dodecahedra that fill the whole space. Mark that the individual dodecahedron is not a regular solid—whereas in the corresponding two-dimensional problem a regular hexagon resulted! The bees' cell consists of the lower half of such a dodecahedron with the six vertical sides so prolonged as to form a hexagonal prism with an open end. Much has been written on this question of the geometry of the honeycomb. The bee's strange social habits and geometric talents could not fail to attract the attention and excite the admiration of their human observers and exploiters. "My house," says the bee in the *Arabian Nights*, "is constructed according to the laws of a most severe architecture; and Euclid himself could learn from studying the geometry of my cells." Maraldi

[1] The arrangement is uniquely determined only if one requires the centers to form a lattice. For the definition of a lattice see p. 96; for a fuller discussion of the problem: D. Hilbert and S. Cohn-Vossen, *Anschauliche Geometrie*, Berlin, 1932, pp. 40–41; and H. Minkowski, *Diophantische Approximationen* Leipzig, 1907, pp. 105–111.

in 1712 seems to be the first to have carried out fairly exact measurements, and he found that the three bottom rhombs of the cell have an obtuse angle α of about 110° and that the angle β they form with the prism walls has the same value. He asked himself the geometric question what the angle α of the rhomb has to be so as to coincide exactly with the latter angle β. He finds $\alpha = \beta = $ 109° 28' and thus assumes that the bees had solved this geometric problem. When principles of minimum were introduced into the study of curves and into mechanics, the idea was not farfetched that the value of α is determined by the most economical use of wax; with every other angle more wax would be needed to form cells of the same volume. This conjecture of Réaumur was confirmed by the Swiss mathematician Samuel Koenig. Somehow Koenig took Maraldi's theoretical value for the one he had actually measured, and finding that his own theoretical value based on the minimum principle deviated from it by 2' (owing to an error of the tables he used in computing $\sqrt{2}$) he concluded that the bees commit an error of less than 2' in solving this minimum problem, of which he says that it lies beyond the reach of classical geometry and requires the methods of Newton and Leibniz. The ensuing discussion in the French Academy was summed up by Fontenelle as Secrétaire perpétuel in a famous judgment in which he denied to the bees the geometric intelligence of a Newton and Leibniz but concluded that in using the highest mathematics they obeyed divine guidance and command. In truth the cells are not as regular as Koenig assumed, it would be difficult to measure the angles even within a few degrees. But more than a hundred years later Darwin still spoke of the bees' architecture as "the most wonderful of known instincts" and adds: "Beyond this stage of perfection in architecture natural selection (which now has replaced divine guidance!) could not lead; for the comb of the hive-bee, as far as we can see, is absolutely perfect in economizing labor and wax."

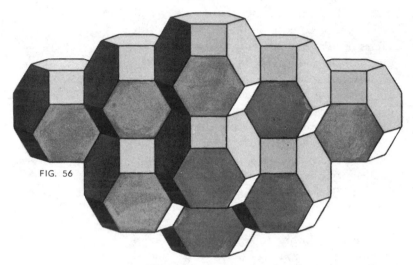

FIG. 56

When one truncates the six corners of an octahedron in a suitable symmetric fashion one obtains a polyhedron bounded by 6 squares and 8 hexagons. This tetrakaideka-hedron was known to Archimedes and rediscovered by the Russian crystallographer Fedorow. Copies of this solid obtained by suitable translations are capable of filling the whole space without overlappings and gaps, just as the rhombic dodecahedron does (Fig. 56). In his Baltimore lectures Lord Kelvin showed how its faces have to be warped and edges curved to fulfill the condition of minimal area. If this is done the partition of space into equal and parallel tetrakaidekahedra gives even a better economy of surface in relation to volume than the plane-faced rhombic dodecahedra. I am inclined to believe that Lord Kelvin's configuration gives the absolute minimum; but so far as I know, this has never been proved.

Let us now return from the three-dimensional space to the two-dimensional plane

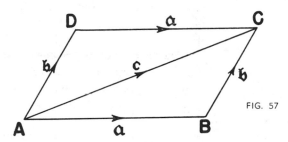

FIG. 57

and engage in a more systematic investigation of symmetry with double infinite rapport. First we have to make this notion precise. As was mentioned before, the translations, the parallel displacements of a plane form a group. A translation $\mathfrak{a}$ can be completely described by fixing the point A' into which it carries a given point A. The translation or vector $\overrightarrow{BB'}$ is the same as the translation $\overrightarrow{AA'}$ if $\overrightarrow{BB'}$ is parallel to $\overrightarrow{AA'}$ and of the same length. The composition of translations is usually denoted by the sign $+$. Thus $\mathfrak{a} + \mathfrak{b}$ is the translation resulting from first carrying out $\mathfrak{a}$ and then $\mathfrak{b}$. If $\mathfrak{a}$ carries the point A into B and $\mathfrak{b}$ carries B into C then $\mathfrak{c} = \mathfrak{a} + \mathfrak{b}$ carries A into C and may thus be indicated by the diagonal vector $\overrightarrow{AC}$ in the parallelogram $ABCD$. Since here $\overrightarrow{AD} = \overrightarrow{BC} = \mathfrak{b}$ and $\overrightarrow{DC} = \overrightarrow{AB} = \mathfrak{a}$ (Fig. 57), we have the commutative law $\mathfrak{a} + \mathfrak{b} = \mathfrak{b} + \mathfrak{a}$ for the composition of translations, or, as one also says, for the addition of vectors. This addition of vectors is nothing else but the law by which two forces $\mathfrak{a}$, $\mathfrak{b}$ unite to form a resultant $\mathfrak{a} + \mathfrak{b} = \mathfrak{c}$ according to the so-called parallelogram of forces. We have the identity or null vector $\mathfrak{o}$ which carries every point into itself, and every translation $\mathfrak{a}$ has its inverse $-\mathfrak{a}$ such that $\mathfrak{a} + (-\mathfrak{a}) = \mathfrak{o}$. It is obvious what $2\mathfrak{a}$, $3\mathfrak{a}$, $4\mathfrak{a}$, . . . stand for, namely $\mathfrak{a} + \mathfrak{a}$, $\mathfrak{a} + \mathfrak{a} + \mathfrak{a}$, $\mathfrak{a} + \mathfrak{a} + \mathfrak{a} + \mathfrak{a}$, etc. The general rule by which the multiple $n\mathfrak{a}$ is defined for every integer n, positive, zero, or negative, is expressed in the formulas

$$(n + 1)\mathfrak{a} = (n\mathfrak{a}) + \mathfrak{a} \qquad \text{and} \qquad 0\mathfrak{a} = \mathfrak{o}.$$

The vector $\mathfrak{b} = \tfrac{1}{3}\mathfrak{a}$ is the unique solution of the equation $3\mathfrak{b} = \mathfrak{a}$. Hence it is clear what $\lambda\mathfrak{a}$ means if λ is a fraction m/n with integral numerator m and denominator n, as for instance $\tfrac{2}{3}$ or $-\tfrac{6}{13}$; and then also by continuity what it means for any real number λ, whether rational or irrational. Two vectors $\mathfrak{e}_1$, $\mathfrak{e}_2$ are linearly independent if no linear combination of them $x_1\mathfrak{e}_1 + x_2\mathfrak{e}_2$ is the null vector $\mathfrak{o}$ unless the two real numbers x_1 and x_2 are zero. The plane is two-dimensional because every vector $\mathfrak{x}$ can be represented uniquely as such a linear combination $x_1\mathfrak{e}_1 + x_2\mathfrak{e}_2$ in terms of two fixed linearly independent vectors $\mathfrak{e}_1$, $\mathfrak{e}_2$. The coefficients x_1, x_2 are called the coordinates of $\mathfrak{x}$ with respect to the basis $(\mathfrak{e}_1, \mathfrak{e}_2)$. After fixing a definite point O as origin (and a vector basis $\mathfrak{e}_1$, $\mathfrak{e}_2$) we can ascribe to every point X two coordinates x_1, x_2 by $\overrightarrow{OX} = x_1\mathfrak{e}_1 + x_2\mathfrak{e}_2$, and vice versa these coordinates x_1, x_2 determine the position of X relative to the "coordinate system" $(O, \mathfrak{e}_1, \mathfrak{e}_2)$.

I am sorry that I had to torture you with these elements of analytic geometry. The purpose of this invention of Descartes' is nothing but to give *names* to the points X in a plane by which we can distinguish and recognize them. This has to be done in a systematic way because there are infinitely many of them; and it is the more necessary as points, unlike persons, are all completely alike, and hence we can distinguish them only by attaching labels to them. The names we employ happen to be pairs of numbers (x_1, x_2).

Besides the commutative law, the addition of vectors—as a matter of fact, the composition of any transformations—satisfies the associative law

$$(\mathfrak{a} + \mathfrak{b}) + \mathfrak{c} = \mathfrak{a} + (\mathfrak{b} + \mathfrak{c}).$$

For the multiplication of vectors $\mathfrak{a}$, $\mathfrak{b}$, . . . by real numbers λ, μ, . . . one has the law

$$\lambda(\mu\mathfrak{a}) = (\lambda\mu)\mathfrak{a}$$

and the two distributive laws

$$(\lambda + \mu)\mathfrak{a} = (\lambda\mathfrak{a}) + (\mu\mathfrak{a}),$$
$$\lambda(\mathfrak{a} + \mathfrak{b}) = (\lambda\mathfrak{a}) + (\lambda\mathfrak{b}).$$

One must ask oneself how the coordinates (x_1, x_2) of an arbitrary vector $\mathfrak{x}$ change as one passes from one vector basis $(\mathfrak{e}_1, \mathfrak{e}_2)$ to another $(\mathfrak{e}'_1, \mathfrak{e}'_2)$. The vectors $\mathfrak{e}'_1$, $\mathfrak{e}'_2$ are expressible in terms of $\mathfrak{e}_1$, $\mathfrak{e}_2$, and vice versa:

(1) $\quad \mathfrak{e}'_1 = a_{11}\mathfrak{e}_1 + a_{21}\mathfrak{e}_2, \qquad \mathfrak{e}'_2 = a_{12}\mathfrak{e}_1 + a_{22}\mathfrak{e}_2$

and

(1') $\quad \mathfrak{e}_1 = a'_{11}\mathfrak{e}'_1 + a'_{21}\mathfrak{e}'_2, \qquad \mathfrak{e}_2 = a'_{12}\mathfrak{e}'_1 + a'_{22}\mathfrak{e}'_2$

Represent the arbitrary vector $\mathfrak{x}$ in terms of the one and the other basis:

$$\mathfrak{x} = x_1\mathfrak{e}_1 + x_2\mathfrak{e}_2 = x'_1\mathfrak{e}'_1 + x'_2\mathfrak{e}'_2.$$

By substituting (1) for e'_1, e'_2 or (1') for e_1, e_2 one finds that the coordinates x_1, x_2 with respect to the first basis are connected with the coordinates x'_1, x'_2 in the second system by the two mutually inverse "homogeneous linear transformations"

$$(2) \quad x_1 = a_{11}x'_1 + a_{12}x'_2, \quad x_2 = a_{21}x'_1 + a_{22}x'_2;$$

$$(2') \quad x'_1 = a'_{11}x_1 + a'_{12}x_2, \quad x'_2 = a'_{21}x_1 + a'_{22}x_2.$$

The coordinates x vary with the vector $\mathfrak{x}$; but the coefficients

$$\begin{pmatrix} a_{11}, & a_{12} \\ a_{21}, & a_{22} \end{pmatrix}, \quad \begin{pmatrix} a'_{11}, & a'_{12} \\ a'_{21}, & a'_{22} \end{pmatrix}$$

are constants. It is easy to see under what circumstances a linear transformation like (2) has an inverse, namely, if and only if its so-called modul $a_{11}a_{22} - a_{12}a_{21}$ is different from 0.

As long as we use no other concepts than those introduced so far, namely (1) addition of vectors $\mathfrak{a} + \mathfrak{b}$, (2) multiplication of a vector $\mathfrak{a}$ by a number λ, (3) the operation by which two points A,B determine the vector $\overrightarrow{AB}$, and concepts logically defined in terms of them, we do affine geometry. In affine geometry any vector basis e_1, e_2 is as good as any other. The notion of the *length* $|\mathfrak{x}|$ of a vector $\mathfrak{x}$ transcends affine geometry and is basic for metric geometry. The square of the length of an arbitrary vector $\mathfrak{x}$ is a quadratic form

$$(3) \qquad g_{11}x_1^2 + 2g_{12}x_1x_2 + g_{22}x_2^2$$

of its coordinates x_1, x_2 with constant coefficients g_{11}, g_{12}, g_{22}. This is the essential content of Pythagoras' theorem. The metric ground form (3) is positive-definite, namely its value is positive for any values of the variables x_1, x_2 except for $x_1 = x_2 = 0$. There exist special coordinate systems, the Cartesian ones, in which this form assumes the simple expression $x_1^2 + x_2^2$; they consist of two mutually perpendicular vectors $\mathfrak{e}_1, \mathfrak{e}_2$ of equal length 1. In metric geometry all Cartesian coordinate systems are equally admissible. Transition from one to the other is affected by an *orthogonal transformation*, i.e. by a homogeneous linear transformation (2), (2′) which leaves the form $x_1^2 + x_2^2$ unaltered, $x_1^2 + x_2^2 = x_1'^2 + x_2'^2$.

But with a slight modification such a transformation may also be interpreted as the algebraic expression of a rotation. If by a rotation around the origin O the Cartesian basis $\mathfrak{e}_1, \mathfrak{e}_2$ passes into the Cartesian basis $\mathfrak{e}_1', \mathfrak{e}_2'$ then the vector $\mathfrak{x} = x_1\mathfrak{e}_1 + x_2\mathfrak{e}_2$ goes into $\mathfrak{x}' = x_1\mathfrak{e}_1' + x_2\mathfrak{e}_2'$, and if you write this as $x_1'\mathfrak{e}_1 + x_2'\mathfrak{e}_2$, using the original basis $(\mathfrak{e}_1, \mathfrak{e}_2)$ as frame of reference throughout, you see that the vector with the coordinates x_1, x_2 goes into that with the coordinates x_1', x_2' where

$$x_1\mathfrak{e}_1' + x_2\mathfrak{e}_2' = x_1'\mathfrak{e}_1 + x_2'\mathfrak{e}_2,$$

and hence

$$(4) \quad x_1' = a_{11}x_1 + a_{12}x_2, \quad x_2' = a_{21}x_1 + a_{22}x_2$$

[formulas (2) with the pairs (x_1, x_2), (x_1', x_2') interchanged].

If vectors are replaced by points the homogeneous linear transformations are replaced throughout by non-homogeneous ones. Let (x_1, x_2), (x_1', x_2') be the coordinates of the same arbitrary point X in two coordinate systems $(O; \mathfrak{e}_1, \mathfrak{e}_2)$, $(O'; \mathfrak{e}_1', \mathfrak{e}_2')$. Then we have

$$\overrightarrow{OX} = x_1\mathfrak{e}_1 + x_2\mathfrak{e}_2, \quad \overrightarrow{O'X} = x_1'\mathfrak{e}_1' + x_2'\mathfrak{e}_2',$$

and since $\overrightarrow{OX} = \overrightarrow{OO'} + \overrightarrow{O'X}$:

$$(5) \qquad x_i = a_{i1}x_1' + a_{i2}x_2' + b_i \quad (i = 1, 2)$$

where we have set $\overrightarrow{OO'} = b_1\mathfrak{e}_1 + b_2\mathfrak{e}_2$. The non-homogeneous differ from homogeneous transformations by the additional terms b_i. The mapping

$$(6) \qquad x_i' = a_{i1}x_1 + a_{i2}x_2 + b_i \quad (i = 1, 2)$$

carrying the point (x_1, x_2) into the point (x_1', x_2') expresses a congruence provided the homogeneous part of the transformation

$$(4) \qquad x_i' = a_{i1}x_1 + a_{i2}x_2 \quad (i = 1, 2),$$

giving the corresponding mapping of the vectors, is orthogonal. (Here of course the coordinates refer to the same fixed coordinate system.) Under these circumstances we call also the non-homogeneous transformation orthogonal. In particular, a *translation* by the vector (b_1, b_2) is expressed by the transformation

$$x_1' = x_1 + b_1, \qquad x_2' = x_2 + b_2.$$

We now return to Leonardo's table of finite rotation groups in the plane,

$$(7) \qquad \begin{cases} C_1, \ C_2, \ C_3, \ \cdots \ ; \\ D_1, \ D_2, \ D_3, \ \cdots \ . \end{cases}$$

The algebraic expression of the operations of one of the groups C_n does not depend on the choice of the Cartesian vector basis. This is not so for the groups D_n; here we normalize the algebraic expression by introducing as the first basic vector $\mathfrak{e}_1$ one that lies in one of the reflection axes. A group of rotations is expressed in terms of the Cartesian coordinate system as a group of orthogonal transformations. Its expressions in any two such coordinate systems linked by an orthogonal transformation are, as we shall say, orthogonally equivalent. Hence what Leonardo had done can now be formulated in algebraic language as follows: He made up a list of groups of orthogonal transformations such that (1) any two of the groups in his list are orthogonally inequivalent, and (2) any finite group of orthogonal transformations is orthogonally equivalent to a group occurring in his list. We say briefly: He made up a *complete* list of *orthogonally inequivalent* finite groups of orthogonal transformations. This seems an unnecessarily involved way of stating a simple situation; but its advantages will presently become evident.

The *symmetry of ornaments* is concerned with discontinuous groups of congruent mappings of the plane. If such a group Δ contains translations it would be absurd to postulate finiteness, for iteration of a translation $\mathfrak{a}$ (different from the identity $\mathfrak{o}$) gives rise to infinitely many translations $n\mathfrak{a}$ ($n = 0, \pm 1, \pm 2, \cdots$). Therefore we replace finiteness by *discontinuity:* it requires that there is no operation in the group arbitrarily close to the identity, except the identity itself. In other words, there is a positive number ϵ such that any transformation (6) in our group for which the numbers

$$\begin{pmatrix} a_{11} - 1, & a_{12}, & b_1 \\ a_{21}, & a_{22} - 1, & b_2 \end{pmatrix}$$

lie between $-\epsilon$ and $+\epsilon$ is the identity (for which all these numbers are zero). The translations contained in our group form a discontinuous group Λ of translations. For such a group there are three possibilities: Either it consists of nothing but the identity, the null vector $\mathfrak{o}$; or all the translations in the group are iterations $x\mathfrak{e}$ of one basic translation $\mathfrak{e} \neq \mathfrak{o}$ ($x = 0, \pm 1, \pm 2, \cdots$); or these translations (vectors) form a two-dimensional *lattice*, namely consist of the linear combinations $x_1\mathfrak{e}_1 + x_2\mathfrak{e}_2$ by *integral* coefficients x_1, x_2 of two linearly independent vectors $\mathfrak{e}_1$, $\mathfrak{e}_2$. The third case is that of double infinite rapport in which we are interested. Here the vectors $\mathfrak{e}_1, \mathfrak{e}_2$ form what we call a *lattice basis*. Choose a point O as origin; those points into which O goes by all the translations of the lattice form a parallelogramatic lattice of points (Fig. 58).

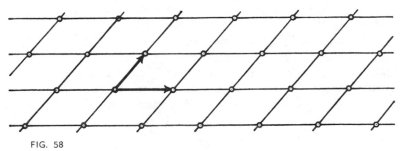

FIG. 58

To what extent, we will ask at once, is the choice of the lattice basis for a given lattice arbitrary? If $\mathfrak{e}'_1, \mathfrak{e}'_2$ is another such basis we must have

(1) $\mathfrak{e}'_1 = a_{11}\mathfrak{e}_1 + a_{21}\mathfrak{e}_2$, $\mathfrak{e}'_2 = a_{12}\mathfrak{e}_1 + a_{22}\mathfrak{e}_2$

where the a_{ij} are *integers*. But also the coefficients in the inverse transformation (1') must be integers, otherwise $\mathfrak{e}'_1$, $\mathfrak{e}'_2$ would not constitute a lattice basis. For the coordinates one gets two mutually inverse linear transformations (2), (2') with integral coefficients

(2'') $\begin{pmatrix} a_{11}, & a_{12} \\ a_{21}, & a_{22} \end{pmatrix}$ and $\begin{pmatrix} a'_{11}, & a'_{12} \\ a'_{21}, & a'_{22} \end{pmatrix}$.

A homogeneous linear transformation with

100

integral coefficients that has an inverse of the same type is called *unimodular* by the mathematicians; one easily sees that a linear transformation with integral coefficients is unimodular if and only if its modul $a_{11}a_{22} - a_{12}a_{21}$ equals $+1$ or -1.

In order to determine all possible discontinuous groups of congruences with double infinite rapport we now proceed as follows. We choose a point O as origin and represent the translations in our group Δ by the lattice L of points into which they carry the point O. Any operation of our group may be considered as a rotation around O followed by a translation. The first, the rotary part, then carries the lattice into itself. Moreover these rotary parts form a discontinuous, and hence *finite* group of rotations $\Gamma = \{\Delta\}$. In the terminology of the crystallographers it is this group which determines the symmetry *class* of the ornament. Γ must be one of the groups in Leonardo's table (7),

(8) $C_n, D_n \quad (n = 1, 2, 3, \cdots),$

but one whose operations carry the lattice L into itself. This relationship between the rotation group Γ and the lattice L imposes certain restrictions on both of them.

As far as Γ is concerned, it excludes from the table all values of n except $n = 1, 2, 3, 4, 6$. Notice that $n = 5$ is among the excluded values! Since the lattice permits the rotation by $180°$, the smallest rotation leaving it invariant must be an aliquot part of $180°$, or of the form

360° divided by 2 or 4 or 6 or 8 or $\cdots$.

We must show that the numbers from 8 on are impossible. Take the case of $n = 8$ and let A among all lattice points $\neq O$ be one that

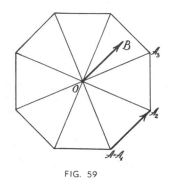

is nearest to O (Fig. 59). Then the whole octagon $A = A_1, A_2, A_3, \cdots$ which arises from A by rotating the plane around O through $\frac{1}{8}$ of the full angle again and again consists of lattice points. Since $\overrightarrow{OA_1}$, $\overrightarrow{OA_2}$ are lattice vectors their difference, the vector $\overrightarrow{A_1A_2}$, must also belong to our lattice, or the point B determined by $\overrightarrow{OB} = \overrightarrow{A_1A_2}$ should be a lattice point. However this leads to a contradiction, since B is nearer to O than $A = A_1$; indeed the side A_1A_2 of the regular octagon is smaller than its radius OA_1. Hence for the group Γ we have only these 10 possibilities:

FIG. 59

(9) $C_1, C_2, C_3, C_4, C_6;$ D_1, D_2, D_3, D_4, D_6.

It is easy to see that for each of these groups there actually exist lattices left invariant by the operations of the group.

Clearly, for C_1 and C_2 any lattice will do, for any lattice is invariant under the identity and the rotation by 180°. But let us consider D_1, which consists of the identity and the reflection in an axis l through O. There are two types of lattices left invariant by this group: the rectangular and the diamond lattices (Fig. 60). The rectangular lattice is obtained by dividing the plane into equal rectangles along lines parallel and perpendicular to l. The corners of the rectangles are the lattice points. A natural basis for the lattice consists of the two sides e_1, e_2 issuing from O of the fundamental rectangle whose left-lower corner is O. The diamond lattice consists of equal rhombs into which the plane is divided by the diagonal lines of the rectangular lattice. The two sides of the fundamental rhomb the left corner of which

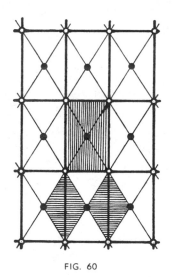

FIG. 60

is O may serve as lattice basis. The lattice points are the corners ○ and the centers ● of the rectangles. (It was an arrangement of trees in such a lozenge lattice which Thomas Browne called quincuncial, thinking of the quincunx ⁝•⁝ as its elementary figure, although the lattice in fact has nothing to do with the number 5.) Shape and size of the fundamental rectangle or rhomb are arbitrary.

After having found the 10 possible groups Γ of rotations and the lattices L left invariant by each of them, one has to paste together a Γ with a corresponding L so as to obtain the full group of congruent mappings. Closer investigation has shown that, while there are 10 possibilities for Γ, there are exactly 17 essentially different possibilities for the full group of congruences Δ. Thus there are 17 essentially different kinds of symmetry possible for a two-dimensional ornament with double infinite rapport. Examples for all 17 groups of symmetry are found among the decorative patterns of antiquity, in particular among the Egyptian ornaments. One can hardly overestimate the depth of geometric imagination and inventiveness reflected in these patterns. Their construction is far from being mathematically trivial. The art of ornament contains in implicit form the oldest piece of higher mathematics known to us. To be sure, the conceptual means for a complete abstract formulation of the underlying problem, namely the mathematical notion of a group of transformations, was not provided before the nineteenth century; and only on this basis is one able to prove that the 17 symmetries already implicitly known to the

Egyptian craftsmen exhaust all possibilities. Strangely enough the proof was carried out only as late as 1924 by George Pólya, now teaching at Stanford.[2] The Arabs fumbled around much with the number 5, but they were of course never able honestly to insert a central symmetry of 5 in their ornamental designs of double infinite rapport. They tried various deceptive compromises, however. One might say that they proved experimentally the impossibility of a pentagon in an ornament.

Whereas it was clear what we meant by saying that there are no other groups of rotations associable with an invariant lattice than the 10 groups (9), our assertion of the existence of not more than 17 different ornamental groups calls for some explanation. For instance, if $\Gamma = C_1$ then the group Δ is one consisting of translations only; but any lattice is possible here, the fundamental parallelogram spanned by the two basic vectors of the lattice may be of any shape and size, we have the choice in a continuously infinite manifold of possibilities. In arriving at the number 17 we count all these as *one* case only; but with what right? Here I need my analytic geometry. If we look at our plane in the light of affine geometry it bears two structures: (i) the metric structure by which every vector $\mathfrak{x}$ has a length the square of which is expressed in terms of coordinates by a positive definite quadratic form (3), the metric ground form; (ii) a lattice structure, owing to the fact that the ornament endows the plane with a vector lattice. In the usual

[2] Cf. his paper "Ueber die Analogie der Kristallsymmetrie in der Ebene," *Zeitschr. f. Kristallographie* *60*, pp. 278–282.

procedure one takes first the metric structure into account and thus introduces the Cartesian coordinate systems in terms of which the metric ground form has a unique normalized expression $x_1^2 + x_2^2$, while in the algebraic representation of the continuous manifold of invariant lattices there remains a variable element. But instead of adapting our coordinates to the metric by admitting Cartesian coordinates only, we could put the lattice structure first and adapt the coordinates to the lattice by choosing e_1, e_2 as a lattice basis, with the effect that the lattice is now normalized in a uniquely determined manner when expressed in terms of the corresponding coordinates x_1, x_2. Indeed the lattice vectors are now those whose coordinates are *integers*. In general we cannot do both at the same time: have a coordinate system in which the metric ground form appears in the normalized form $x_1^2 + x_2^2$, and the lattice consists of all vectors with integral coordinates x_1, x_2. We now follow the second procedure, which turns out to be mathematically more advantageous. I consider this analysis as one which is of basic importance for all morphology.

As an example consider once more D_1. If the invariant lattice is rectangular and the lattice basis is chosen in the natural manner described above, then D_1 consists of the identity and the operation

$$x_1' = x_1, \qquad x_2' = -x_2.$$

The metric ground form can be any positive form of the special type $a_1 x_1^2 + a_2 x_2^2$. If the invariant lattice is a diamond lattice and the sides of the fundamental diamond are chosen as the lattice basis, then D_1 contains the identity and the further operation

$$x_1' = x_2, \qquad x_2' = x_1.$$

The metric ground form may be any positive form of the special type $a(x_1^2 + x_2^2) + 2bx_1x_2$. But instead of D_1 we now obtain two groups D_1^a, D_1^b of linear transformations with integral coefficients, which, though orthogonally, are no longer unimodularly equivalent, the one consisting of the two operations with the coefficient matrices

$$\begin{pmatrix} 1 & 0 \\ 0 & 1 \end{pmatrix}, \qquad \begin{pmatrix} 1 & 0 \\ 0 & -1 \end{pmatrix},$$

the other of

$$\begin{pmatrix} 1 & 0 \\ 0 & 1 \end{pmatrix}, \qquad \begin{pmatrix} 0 & 1 \\ 1 & 0 \end{pmatrix}.$$

Two groups of homogeneous linear transformations are, of course, called unimodularly equivalent if they both represent the same group of operations, the one in terms of one, the other in terms of another lattice basis, i.e. if they change into each other by a unimodular transformation of coordinates.

In the lattice-adapted coordinate system the operations of Γ now appear as homogeneous linear transformations (4) with integral coefficients a_{ij}; for as each carries the lattice into itself, x_1', x_2' assume integral values whenever integral values are assigned to x_1 and x_2. The arbitrariness in the choice of the lattice basis finds its expression by the agreement to consider unimodularly equivalent groups of linear transformations as the same thing. Besides having integral coefficients the transformations of Γ will leave a certain positive definite quadratic form (3) invariant. But this is really no additional restriction; indeed it can be shown that for any finite group of linear transformations with real coefficients one may construct positive quadratic forms left invariant under these transformations.[3] How many different, i.e. unimodularly inequivalent, finite groups of linear transformations with integral coefficients in two variables, exist then? Ten, namely our old friends (9)? No, there are more, since we have seen that D_1, for instance, breaks up into two inequivalent cases D_1^a, D_1^b. The same happens to D_2 and D_3, with the result that there are exactly 13 unimodularly inequivalent finite groups of linear operations with integral coefficients. From a mathematical standpoint this is the really interesting result rather than the table (9) of the 10 groups of rotations with invariant lattices.

In a last step one has to introduce the translational parts of the operations, and one obtains 17 unimodularly inequivalent discontinuous groups of non-homogeneous linear transformations which contain all the translations

$$x_1' = x_1 + b_1, \qquad x_2' = x_2 + b_2$$

with integral b_1, b_2 and no other translations. This last step offers little difficulty, and the remarks that remain to be made are better based on the 13 finite groups Γ of homogeneous transformations which result from cancelling the translatory parts.

So far only the lattice structure of the plane has been taken into account. Of course, the metric of the plane cannot be ignored forever. And it is here that the continuous

[3] This is a fundamental theorem due to H. Maschke. The proof is simple enough: Take any positive quadratic form, e.g. $x_1^2 + x_2^2$, perform on it each of the transformations S of our group and add the forms thus obtained: the result is an invariant positive form.

aspect of the problem comes in. For each of the 13 groups Γ there exist invariant positive quadratic forms

$$G(x) = g_{11}x_1^2 + 2g_{12}x_1x_2 + g_{22}x_2^2.$$

Such a form is characterized by its coefficients (g_{11}, g_{12}, g_{22}). The form $G(x)$ is not uniquely determined by Γ; for instance $G(x)$ may be replaced by any multiple $c \cdot G(x)$ with a real positive constant factor c. All positive quadratic forms $G(x)$ left invariant by the operations of Γ form a continuous convex "cone" of simple nature and of 1, 2, or 3 dimensions. For instance in the cases D_1^a and D_1^b we had the two-dimensional manifolds of all positive forms of the types $a_1x_1^2 + a_2x_2^2$ and $a(x_1^2 + x_2^2) + 2bx_1x_2$ respectively. The metric ground form is always one in the manifold of invariant forms.

In the full description of the ornamental groups Δ we have now clearly divided those features which are discrete, and those capable of varying over a continuous manifold. The discrete feature is exhibited by representing the group in terms of lattice-adapted coordinates and turns out to be one of 17 definite distinct groups. To each of them there corresponds a continuum of possibilities for the metric ground form $G(x)$, from which the one actual metric ground form must be picked. The advantage of adapting the coordinate system to the lattice rather than to the metric becomes visible in the fact that now the variable element $G(x)$ varies over one simple convex continuous manifold, while in terms of the metric-adapted coordinates the lattice L, which then appears as the variable element, ranges over a continuum that may consist of several parts, as the example of D_1 showed. The advantage is fully revealed only if one passes from the truncated homogeneous group $\Gamma = \{\Delta\}$ to the full ornamental group Δ. The splitting into something discrete and something continuous seems to me a basic issue in all morphology, and the morphology of ornaments and crystals establishes a paragon by the clearcut way in which this distinction is carried out.

After all these somewhat abstract mathematical generalities I am now going to show you a few pictures of surface ornaments with double infinite rapport. You find them on wall papers, carpets, tiled floors, parquets, all sorts of dress material, especially prints, and so forth. Once one's eyes are opened, one will be surprised by the numerous symmetric patterns which surround us in our daily lives. The greatest masters of the geometric art of ornament were the Arabs. The wealth of stucco ornaments decorating the walls of such buildings of Arabic origin as the Alhambra in Granada is simply overwhelming.

For the purpose of description it is good to know what a congruent mapping in two dimensions looks like. A proper motion may be either a translation or a rotation around a point O. If such a rotation occurs in our symmetry group and all the rotations around O occurring in it are multiples of the rotation by $360°/n$, we call O a pole of multiplicity n or simply n-pole. We know that no other values except $n = 2, 3, 4, 6$ are possible. An improper congruence is either a reflection in a line l, or such a reflection combined with a translation $\mathfrak{a}$ along l. If it occurs in our group, l is called an axis or gliding axis respectively. In the latter case iteration of the congruence leads to the translation by the

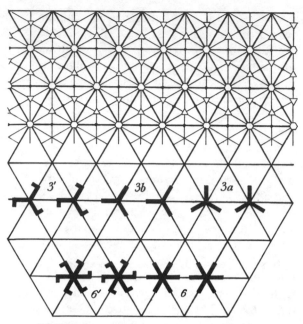

FIG. 61

vector $2\mathfrak{a}$; hence the gliding vector $\mathfrak{a}$ must be one-half of a lattice vector of our group.

The first picture (Fig. 61) is a drawing of the hexagonal lattice, the discussion of which started today's lecture. It has a very rich symmetry. There are poles of multiplicities 2, 3, and 6, indicated in the diagram by dots, small triangles, and hexagons respectively. The vectors joining two 6-poles are the lattice vectors. The lines in the figure are the axes. There are also gliding axes, which have not been shown in the drawing; they run mid-way between and parallel to the axes. The possible symmetry groups of the hexagonal type are five in number and are obtained by putting one of the simple figures **6** or **6'** or **3'** or **3a** or **3b** into each of the six-poles. Designs **6** and **6'** preserve the multiplicity 6 of these poles, but **6'** destroys the symmetry axes. Designs $3'$, $3a$, and $3b$ reduce the multiplicity of those poles to 3; among them $3'$ is without symmetry axes, in $3a$ axes pass through every 3-pole, in $3b$ only through those (one-third of the whole number) which had been six-fold before. The homogeneous groups are D_6, C_6, C_3, D_3^a, D_3^b respectively, where D_3^a, D_3^b are the two unimodularly inequivalent forms assumed by D_3 in a lattice-adapted coordinate system.

There now follow some actual ornaments of Moorish, Egyptian, and Chinese origin. This window of a mosque in Cairo, of the fourteenth century (Fig. 62), is of the hexa-

FIG. 62

gonal class D_6. The elementary figure is a
trefoil knot the various units of which are
interlaced with superb artistry. Almost un-
interrupted tracks cross the design in the
three directions arising from the horizontal
by rotations through 0°, 60°, 120°; the mid-
lines of these tracks are gliding axes. You
can easily discover lines that are ordinary
axes. Such axes are absent from this
Azulejos ornament (Fig. 63) decorating the
back of the alcove in the Sala de Camas of the

FIG. 63

Alhambra in Granada. The group is **3'** or **6'**, according to whether or not one takes account of the colors. This is one of the finer tricks of the ornamental art that the symmetry of the geometric pattern as expressed by a certain group Δ is reduced by the coloring to a lower symmetry expressed by a subgroup of Δ. A symmetry of the square class D_4 is exhibited by this (Fig. 64) well-known design for brick pavements; the amusing thing about it is that no ordi-

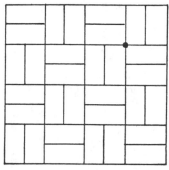

FIG. 64

FIG. 65

nary axes, only gliding axes, pass through the 4-poles (one of which is marked). Of the same symmetry is the Egyptian ornament shown next (Fig. 65), as well as the two following Moorish ornaments (Fig. 66). A monumental work on our subject is Owen Jones' *Grammar of ornaments*, from which some of these illustrations are taken. Of a more special character is the *Grammar of Chinese lattice* by Daniel Sheets Dye, which deals with the lattice work the Chinese use for the support of their paper windows. I reproduce here (Figs. 67 and 68) two characteristic designs from that volume, one of the hexagonal and one of the D_4-type.

113

FIG. 66

FIG. 67

114

FIG. 68

I wish I could analyze some of these ornaments in detail. But a prerequisite for such
an investigation would be an explicit algebraic description of the 17 ornamental
groups. What this lecture aimed at was
more a clarification of the general mathematical principles underlying the morphology
of ornaments (and crystals) than a group-
theoretic analysis of individual ornaments.
Shortness of time has prevented me from
doing justice to both sides, the abstract and
the concrete. I tried to explain the basic
mathematical ideas, and I showed you some
pictures: the bridge between them I indicated, but I could not lead you over it step
by step.

CRYSTALS
THE GENERAL MATHEMATICAL
IDEA OF SYMMETRY

CRYSTALS
THE GENERAL MATHEMATICAL
IDEA OF SYMMETRY

IN THE LAST LECTURE we considered for two dimensions the problem of making up a complete list (i) of all orthogonally inequivalent finite groups of homogeneous orthogonal transformations, (ii) of all such groups as have invariant lattices, (iii) of all unimodularly inequivalent finite groups of homogenous transformations with integral coefficients, (iv) of all unimodularly inequivalent discontinuous groups of non-homogeneous linear transformations which contain the translations with integral coordinates but no other translations.

Problem (i) was answered by Leonardo's list

$$C_n, D_n \quad (n = 1, 2, 3, \cdots),$$

(ii) by limiting the index n in it to the values $n - 1, 2, 3, 4, 6$. The numbers h_{I}, h_{II}, h_{III}, h_{IV} of groups in these four lists turned out to be

$$\infty, 10, 13, 17$$

respectively. The most important problem is doubtless (iii). One could have posed these same four questions for the one-dimensional line rather than the two-dimensional plane. There the answer would have been very simple, and one would have found all the numbers h_{I}, h_{II}, h_{III}, h_{IV} equal to 2. Indeed in each of the cases (i), (ii), (iii) the group consists either of the identity $x' = x$ alone, or of the identity and the reflection $x' = -x$.

But what we are intending now is not to descend from 2 to 1, but ascend from 2 to 3 dimensions. All finite groups of rotations in 3 dimensions were listed at the end of the second lecture; I repeat them here:

List A:

$$C_n, \qquad \bar{C}_n, \qquad C_{2n}C_n \qquad (n = 1, 2, 3, \cdots)$$

$$D'_n, \qquad \bar{D}'_n, \qquad D'_{2n}D'_n, \qquad D'_nC_n$$
$$(n = 2, 3, \cdots)$$

$$T, \quad W, \quad P; \qquad \bar{T}, \quad \bar{W}, \quad \bar{P}; \qquad WT.$$

If one requires that the operations of the group leave a lattice invariant, only axes of rotation of multiplicity 2, 3, 4, 6 are admissible. And by this restriction our table reduces to the following

List B:

$$C_1, \ C_2, \ C_3, \ C_4, \ C_6; \qquad \bar{C}_1, \ \bar{C}_2, \ \bar{C}_3, \ \bar{C}_4, \ \bar{C}_6;$$
$$D'_2, \ D'_3, \ D'_4, \ D'_6; \qquad \bar{D}'_2, \ \bar{D}'_3, \ \bar{D}'_4, \ \bar{D}'_6;$$
$$C_2C_1, \ C_4C_2, \ C_6C_3;$$
$$D'_4D'_2, \ D'_6D'_3;$$
$$D'_2C_2, \ D'_3C_3, \ D'_4C_4, \ D'_6C_6;$$
$$T, \ W, \ \bar{T}, \ \bar{W}, \ WT.$$

It contains 32 members. One easily convinces oneself that each of these 32 groups possesses invariant lattices. In three dimensions the numbers h_I, h_{II}, h_{III}, h_{IV} have the values

$$\infty, \ 32, \ 70, \ 230.$$

In its algebraic formulation our problem may be posed for any number m of variables, $x_1, x_2, \cdots, x_m$ instead of just 2 or 3, and the corresponding theorems of finiteness have been proved. The methods are of the greatest mathematical interest. The combination "metric plus lattice" lies at the bottom of the arithmetical theory of quadratic forms which, inaugurated by Gauss, played a central part in the theory of numbers throughout the nineteenth century. Dirichlet, Hermite, and more recently Minkowski and Siegel have contributed to this line of research. The investigation of ornamental symmetry in m dimensions is based on the results won by these authors and on the algebraic and the more refined arithmetic theory of so-called algebras or hypercomplex number systems, on which the last generation of algebraists, in this country L. Dickson above all, have spent much effort.

We decorate surfaces with flat ornaments; art has never gone in for solid ornaments. But they are found in nature. The arrangements of atoms in a crystal are such patterns. The geometric forms of crystals with their plane surfaces are an intriguing phenomenon of nature. However the real physical symmetry of a crystal is not so much shown by its outward appearance as by the inner physical structure of the crystalline substance. Let us suppose that this substance fills the entire space. Its macroscopic symmetry will find its expression in a group Γ of rotations. Only such orientations of the crystal in space are physically indistinguishable as are carried into each other by a rotation of this group. For example, light, which in general propagates with different velocities in different directions in the crystalline medium, will propagate with the same speed in any two directions that arise from each other by a rotation of the group Γ. So for all other physical properties. For an isotropic medium the group Γ consists of all rotations, but for a crystal it is made up of a finite number of rotations, sometimes even of nothing but the identity. Early in the history of crystallography the law of rational indices was derived from the arrangement of the plane surfaces of crystals. It led to the hypothesis of the lattice-like atomic structure of crystals. This hypothesis, which explains the law of rational indices, has now been definitely confirmed by the Laue interference patterns, which are essentially X-ray photographs of crystals.

More exactly the hypothesis states that the discontinuous group Δ of congruences which carry the arrangement of atoms in our crystal into itself contains the maximum number 3 of linearly independent translations. Incidentally this hypothesis can be reduced to much simpler requirements. Atoms which go over into each other by an operation of Δ may be called equivalent. The equivalent atoms form a regular point-set in the sense that the set is carried into itself by every operation of Δ and that for any two points in the set there is an operation of Δ which carries the one into the other. Speaking of the arrangements of atoms I refer to their positions in equilibrium; in fact the atoms oscillate around these positions. Perhaps one should take a lead from quantum mechanics and substitute for the atoms' exact positions their average distribution density: this density function in space is invariant with respect to the operations of Δ. The group $\Gamma = \{\Delta\}$ of the rotational parts of those congruences that are members of Δ leaves the lattice L of points invariant which arise from the origin O by the translations contained in Δ. The resulting 32 possibilities for Γ enumerated in List B correspond to the 32 existing symmetry classes of crystals. For the group Δ itself we have 230 distinct possibilities, as was mentioned above.[1]

[1] See for instance P. Niggli, *Geometrische Kristallographie des Diskontinuums*, Berlin, 1920.

While $\Gamma = \{\Delta\}$ describes the manifest macroscopic spatial and physical symmetry, Δ defines the microscopic atomic symmetry hidden behind it. You probably all know on what the success of von Laue's photography of crystals depends. The image of an object traced by light of a certain wave length will be fairly accurate only with respect to details of considerably greater dimensions than that wave length, whereas details of smaller dimensions are leveled down. Now the wave length of ordinary light is about a thousand times as big as the atomic distances. However X-rays are light rays the wave length of which is exactly of the desirable order 10^{-8} centimeters. In this way von Laue killed two birds with one stone: he confirmed the lattice structure of crystals, and proved what had been merely a tentative hypothesis at the time of his discovery (1912), that X-rays consist of shortwave light. Even so, the portraits of the atomic pattern which his diagrams show are not likenesses in the literal sense. By observing a slit whose width is only a few wave lengths, you obtain a somewhat contorted image of the slit made up by interference fringes. In the same sense these Laue diagrams are interference patterns of the atomic lattice. Yet one is able to compute from such photographs the actual arrangement of atoms, the scale being set by the wave length of the illuminating X-rays. Here are two Laue diagrams (Figs. 69 and 70), both of zinc-blende from Laue's original paper (1912); the pictures are taken in such directions as to exhibit the symmetry around an axis of order 4 and 3 respectively. Whereas in the oral lecture I could show various three-dimensional (enlarged) models of the actual arrangement of atoms, a photo-

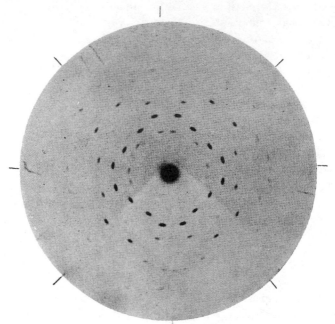

FIG. 69

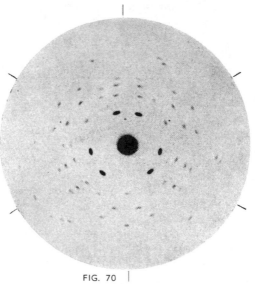

FIG. 70

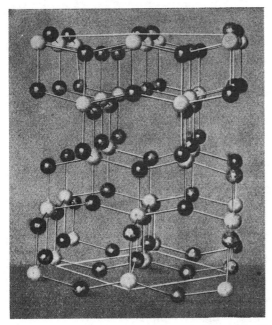

FIG. 71

graph of one such model (Fig. 71) must suffice for this printed text: it represents a small part of an Anatas crystal of the chemical constitution TiO_2; the light balls are the Ti-, the dark ones the O-atoms.

In spite of all contortion which mars our X-ray likenesses, the symmetry of the crystal is faithfully portrayed. This is a special case of the following general principle: If conditions which uniquely determine their effect possess certain symmetries, then the effect will exhibit the same symmetry. Thus Archimedes concluded a priori that equal weights balance in scales of equal arms. Indeed the whole configuration is symmetric with respect to the midplane of the scales, and therefore it is impossible that one mounts while the other sinks. For the same reason

125

we may be sure that in casting dice which are perfect cubes, each side has the same chance, $\frac{1}{6}$. Sometimes we are thus enabled to make predictions a priori on account of symmetry for special cases, while the general case, as for instance the law of equilibrium for scales with arms of different lengths, can only be settled by experience or by physical principles ultimately based on experience. As far as I see, all a priori statements in physics have their origin in symmetry.

To this epistemological remark about symmetry I add a second. The morphological laws of crystals are today understood in terms of atomic dynamics: if equal atoms exert forces upon each other that make possible a definite state of equilibrium for the atomic ensemble, then the atoms in equilibrium will of necessity arrange themselves in a regular system of points. The nature of the atoms composing the crystal determines under given external conditions their metric disposition, for which the purely morphological investigation summed up in the 230 groups of symmetry Δ had still left a continuous range of possibilities. The dynamics of the crystal lattice is also responsible for the crystal's physical behavior, in particular for the manner of its growth, and this in turn determines the peculiar shape it assumes under the influence of environmental factors. No wonder then that crystals actually occurring in nature display the possible types of symmetry in that abundance of different forms at which Hans Castorp on his Magic Mountain marvelled. The visible characteristics of physical objects usually are the results of constitution and environment. Whether water, whose molecule has a definite chemical constitution, is solid, liquid, or vaporous depends on temperature. Temperature is the environmental factor kat' exochen. The examples of crystallography, chemistry, and genetics cause one to suspect that this duality, described by the biologists as that of genotype and phenotype or of nature and nurture, is in some way bound up with the distinction between discrete and continuous; and we have seen how such a splitting into the discrete and the continuous can be carried out for the characteristic features of crystals in a most convincing way. But I will not deny that the general problem is in need of further epistemological clarification.

It is high time for me now to close the discussion of geometric symmetries dwelling in ornaments and crystals. The chief aim of this last lecture is to show the principle of symmetry at work in questions of physics and mathematics of a far more fundamental nature, and to rise from these and its previous applications to a final general statement of the principle itself.

What the *theory of relativity* has to do with symmetry was briefly explained in the first lecture: before one studies geometric forms in space with regard to their symmetry one must examine the structure of space itself under the same aspect. Empty space has a very high grade of symmetry: every point is like any other, and at a point there is no intrinsic difference between the several directions. I told you that Leibniz had given the geometric notion of similarity this philosophical twist: Similar, he said, are two things which are indiscernible when each is considered by itself. Thus two squares in the same plane may show many differences when one regards their relation to each other;

for instance, the sides of the one may be inclined by 34° against the sides of the other. But if each is taken by itself, any objective statement made about one will hold for the other; in this sense they are indiscernible and hence similar. What requirements an objective statement has to meet I shall illustrate by the meaning of the word "vertical." Contrary to Epicurus we moderns do not consider the statement about a line that it is vertical to be an objective one, because we see in it an abbreviation of the more complete statement that the line has the direction of gravity at a certain point P. Thus the gravitational field enters into the proposition as a contingent factor, and moreover there enters into it an individually exhibited point P on which we lay the finger by a demonstrative act as is expressed in words like I, here, now, this. Hence Epicurus' belief is shattered as soon as it is realized that the direction of gravity is different here at the place where I live and at the place where Stalin lives, and that it can also be changed by a redistribution of matter.

Let these brief remarks suffice here instead of a more thorough analysis of objectivity. In concreto, as far as geometry is concerned, we have followed Helmholtz in adopting as the one basic objective relation in space that of congruence. At the beginning of the second lecture we spoke of the group of congruent transformations, which is contained as a subgroup in the group of all similarities. Before going on I wish to clarify a little further the relationship of these two groups. For there is the disquieting question of the *relativity of length*.

In ordinary geometry length is relative: a building and a small-scale model of it are similar; the dilatations are included among the automorphisms. But physics has revealed that an absolute standard length is built into the constitution of the atom, or rather into that of the elementary particles, in particular the electron with its definite charge and mass. This atomic standard length becomes available for practical measurements through the wave lengths of the spectral lines of the light emitted by atoms. The absolute standard thus derived from nature itself is much better than the conventional standard of the platinum-iridium meter bar kept in the vaults of the Comité International des Poids et Mesures in Paris. I think the real situation has to be described as follows. Relative to a complete system of reference not only the points in space but also all physical quantities can be fixed by numbers. Two systems of reference are equally admissible if in both of them all universal geometric and physical laws of nature have the same algebraic expression. The transformations mediating between such equally admissible systems of reference form the group of *physical automorphisms;* the laws of nature are invariant with respect to the transformations of this group. It is a fact that a transformation of this group is uniquely determined by that part of it that concerns the coordinates of space points. Thus we can speak of the physical automorphisms *of space.* Their group does not include the dilatations, because the atomic laws fix an absolute length, but it contains the reflections because no law of nature indicates an intrinsic difference between left and right. Hence the group of physical automorphisms is the group of all proper and improper congruent mappings. If we call two configurations in space congruent provided they are carried over into each other by a transformation of this group, then bodies which are mirror images of each other are congruent. I think it is necessary to substitute this definition of congruence for that depending on the motion of rigid bodies, for reasons similar to those which induce the physicist to substitute the thermodynamical definition of temperature for an ordinary thermometer. Once the group of physical automorphisms = congruent mappings has been established, one may define geometry as the science dealing with the relation of congruence between spatial figures, and then the *geometric automorphisms* would be those transformations of space which carry any two congruent figures into congruent figures—and one need not be surprised, as Kant was, that this group of geometric automorphisms is wider than that of physical automorphisms and includes the dilatations.

All these considerations are deficient in one respect: they ignore that physical occurrences happen not only in space but in *space and time;* the world is spread out not as a three- but as a four-dimensional continuum. The symmetry, relativity, or homogeneity of this four-dimensional medium was first correctly described by Einstein. Has the statement, we ask, that two events occur at the same place an objective significance? We are inclined to say yes; but it is clear, if we do so, we understand position as position relative to the earth on which we spend our life. But is it sure that the earth rests? Even our youngsters are now told in school that it rotates and that it moves about in space. Newton wrote his treatise *Philosophiae naturalis principia mathematica* to answer this question, to deduce, as he said, the absolute motion of bodies from their differences, the observable relative motions, and from the forces acting upon the bodies. But although he firmly believed in absolute space, i.e. in the objectivity of the statement that two events occur in the same place, he did not succeed in objectively distinguishing rest of a mass point from all other possible motions, but only motion in a straight line with uniform velocity, the so-called uniform translation, from all other motions. Again, has the statement that two events occur at the same time (but at different places, say here and on Sirius) objective meaning? Until Einstein, people said yes. The basis of this conviction is obviously people's habit of considering an event as happening at the moment when they observe it. But the foundation of this belief was long ago shattered by Olaf Roemer's discovery that light propagates not instantaneously but with finite velocity. Thus one came to realize that in the four-dimensional continuum of space-time only the coincidence of two world points, "here-now," or their immediate vicinity has a directly verifiable meaning. But whether a stratification of this four-dimensional continuum in three-dimensional layers of simultaneity and a cross-fibration of one-dimensional fibers, the world-lines of points resting in space, describe objective features of the world's structure became doubtful. What Einstein did was this: without bias he collected all the physical evidence we have about the real structure of the four-dimensional space-time continuum and thus derived its true group of automorphisms. It is called the Lorentz group after the Dutch physicist H. A. Lorentz who, as Einstein's John the

Baptist, prepared the way for the gospel of relativity. It turned out that according to this group there are neither invariant layers of simultaneity nor invariant fibers of rest. The light cone, the locus of all world-points in which a light signal given at a definite world-point O, "here-now," is received, divides the world into future and past, into that part of the world which can still be influenced by what I do at O and the part which cannot. This means that no effect travels faster than light, and that the world has an objective causal structure described by these light cones issuing from every world point O. Here is not the place to write down the Lorentz transformations and to sketch how special relativity theory with its fixed causal and inertial structure gave way to general relativity where these structures have become flexible by their interaction with matter.[2] I only want to point out that it is the inherent symmetry of the four-dimensional continuum of space and time that relativity deals with.

We found that objectivity means invariance with respect to the group of automorphisms. Reality may not always give a clear answer to the question what the actual group of automorphisms is, and for the purpose of some investigations it may be quite useful to replace it by a wider group. For instance in plane geometry we may be interested only in such relations as are invariant under parallel or central projections; this is the origin of affine and projective geometry. The mathematician will prepare for all such

[2] One may compare my recent lecture at the Munich meeting of the Gesellschaft deutscher Naturforscher: "50 Jahre Relativitätstheorie," *Die Naturwissenschaften 38* (1951), pp. 73–83.

eventualities by posing the general problem, how for a given group of transformations to find its invariants (invariant relations, invariant quantities, etc.), and by solving it for the more important special groups— whether these groups are known or are not known to be the groups of automorphisms for certain fields suggested by nature. This is what Felix Klein called "a geometry" in the abstract sense. A geometry, Klein said, is defined by a group of transformations, and investigates everything that is invariant under the transformations of this given group. Of symmetry one speaks with respect to a subgroup γ of the total group. Finite subgroups deserve special attention. A figure, i.e. any point-set, has the peculiar kind of symmetry defined by the subgroup γ if it goes over into itself by the transformations of γ.

The two great events in twentieth century physics are the rise of relativity theory and of *quantum mechanics*. Is there also some connection between quantum mechanics and symmetry? Yes indeed. Symmetry plays a great role in ordering the atomic and molecular spectra, for the understanding of which the principles of quantum physics provide the key. An enormous amount of empirical material concerning the spectral lines, their wave lengths, and the regularities in their arrangement had been collected before quantum physics scored its first success; this success consisted in deriving the law of the so-called Balmer series in the spectrum of the hydrogen atom and in showing how the characteristic constant entering into that law is related to charge and mass of the electron and Planck's famous constant of action h. Since then the interpretation of the spectra has accompanied the development of quantum physics; and the decisive new features, the electronic spin and Pauli's strange exclusion principle, were discovered in this way. It turned out that, once these foundations had been laid, symmetry could be of great help in elucidating the general character of the spectra.

Approximately, an atom is a cloud of electrons, say of n electrons, moving around a fixed nucleus in O. I say approximately, since the assumption that the nucleus is fixed is not exactly true and is even less justified than treatment of the sun as the fixed center of our planetary system. For the mass of the sun is 300,000 times as big as that of an individual planet like Earth, whereas the proton, the nucleus of the hydrogen atom, is less than 2000 times as heavy as the electron. Even so it is a good approximation! We distinguish the n electrons by attaching the labels 1, 2, $\cdots$, n to them; they enter into the laws ruling their motion by the coordinates of their positions P_1, $\cdots$, P_n with respect to a Cartesian coordinate system with origin O. The symmetry prevailing is twofold. First we must have invariance with respect to transition from one Cartesian coordinate system to another; this symmetry comes from the rotational symmetry of space and is expressed by the group of geometric rotations about O. Secondly, all electrons are alike; the distinction by their labels 1, 2, $\cdots$, n is one not by essence, but by name only: two constellations of electrons that arise from each other by an arbitrary permutation of the electrons are indiscernible. A permutation consists of a re-arrangement of the labels; it is really a one-to-one mapping of the set of labels $(1, 2, \cdots, n)$ into itself, or if you will, of the corresponding set of points $(P_1, \cdots, P_n)$. Thus, for example, in case of $n = 5$ electrons the laws must be unaffected if the points P_1, P_2, P_3, P_4, P_5 are replaced by P_3, P_5, P_2, P_1, P_4 [permutation $1 \rightarrow 3, 2 \rightarrow 5, 3 \rightarrow 2, 4 \rightarrow 1, 5 \rightarrow 4$]. The permutations form a group of order $n! = 1 \cdot 2 \cdots n$, and the second kind of symmetry is expressed by this group of permutations. Quantum mechanics represents the state of a physical system by a vector in a space of many, actually of infinitely many, dimensions. Two states that arise from each other, either by a virtual rotation of the system of electrons or by one of their permutations, are connected by a linear transformation associated with that rotation or that permutation. Hence the profoundest and most systematic part of group theory, the theory of representations of a group by linear transformations, comes into play here. I must refrain from giving you a more precise account of this difficult subject. But here symmetry once more has proved the clue to a field of great variety and importance.

From art, from biology, from crystallography and physics I finally turn to *mathematics*, which I must include all the more because the essential concepts, especially that of a group, were first developed from their applications in mathematics, more particularly in the theory of algebraic equations. An algebraist is a man who deals in numbers, but the only operations he is able to perform are the four species $+$, $-$, $\times$, $\div$. The numbers which arise by the four species from 0 and 1 are the rational numbers. The field F of these numbers is closed with respect to the four species, i.e. sum, difference, and product of two rational numbers are rational numbers, and so is their quotient provided the divisor is different from zero. Thus the algebraist would have had no reason to step outside this domain F, had not the demands of geometry and physics forced the mathematicians to engage in the dire business of analysing *continuity* and to embed the rational numbers in the continuum of all real numbers. This necessity first appeared when the Greeks discovered that the diagonal and side of a square are incommensurable. Not long afterwards Eudoxus formulated the general principles on which to base the construction of a system of real numbers suitable for all measurements. Then during the Renaissance the problem of solving algebraic equations led to the introduction of complex numbers $a + bi$ with real components (a,b). The mystery that first shrouded them and their imaginary unit $i = \sqrt{-1}$ was completely dissolved when one recognized that they are nothing but pairs (a, b) of ordinary real numbers, pairs for which addition and multiplication are so defined as to preserve all the familiar laws of arithmetics. This can indeed be done in such a way that any real number a may be identified with the complex number $(a, 0)$ and that the square $i \cdot i = i^2$ of $i = (0, 1)$ equals -1, more explicitly $(-1, 0)$. Thus the equation $x^2 + 1 = 0$ not solvable by any real number x became solvable. At the beginning of the nineteenth century it was proved that the introduction of the complex numbers had made solvable not only this but all algebraic equations: the equation

(1) $f(x) =$
$x^n + a_1 x^{n-1} + a_2 x^{n-2} + \cdots + a_{n-1} x + a_n = 0$

for the unknown x, whatever its degree n and

its coefficients a_ν, has n solutions, or "roots" (as one is wont to say) $\vartheta_1, \vartheta_2, \cdots, \vartheta_n$, so that the polynomial $f(x)$ itself decomposes into the n factors

$$f(x) = (x - \vartheta_1)(x - \vartheta_2) \cdots (x - \vartheta_n).$$

Here x is a variable or indeterminate, and the equation is to be interpreted as stating that the two polynomials on either side coincide coefficient for coefficient.

Such relations between two indeterminate numbers x, y as the algebraist is able to construct with his operations of addition and multiplication can always be brought into the form $R(x, y) = 0$ where the function $R(x, y)$ of the two variables x, y is a polynomial, i.e. a finite sum of monomials of the type

$$a_{\mu,\nu} x^\mu y^\nu \quad (\mu, \nu = 0, 1, 2, \cdots)$$

with rational coefficients $a_{\mu\nu}$. These relations are the "objective relations" accessible to him. Given two complex numbers α, β he will therefore ask what polynomials $R(x, y)$ with rational coefficients exist which get annulled by substituting the value α for the indeterminate x and β for y. From two one may pass to any number of given complex numbers $\vartheta_1, \cdots, \vartheta_n$. The algebraist will ask for the automorphisms of this set Σ of numbers, namely for those permutations of $\vartheta_1, \cdots, \vartheta_n$ which destroy none of the algebraic relations $R(\vartheta_1, \cdots, \vartheta_n) = 0$ existing between them. Here $R(x_1, \cdots, x_n)$ is any polynomial with rational coefficients of the n indeterminates $x_1, \cdots, x_n$ which is annulled by substituting the values $\vartheta_1, \cdots, \vartheta_n$ for $x_1, \cdots, x_n$. The automorphisms form a group which is called the *Galois group*, after the French mathematician

Evariste Galois (1811–1832). As this description shows, Galois' theory is nothing else but the relativity theory for the set Σ, a set which, by its discrete and finite character, is conceptually so much simpler than the infinite set of points in space or space-time dealt with by ordinary relativity theory. We stay entirely within the confines of algebra when we assume in particular that the members $\vartheta_1, \cdots, \vartheta_n$ of the set Σ are defined as the n roots of an algebraic equation (1), $f(x) = 0$, of nth degree, with rational coefficients a_ν. One then speaks of the Galois group of the equation $f(x) = 0$. It may be difficult enough to determine the group, requiring as it does a survey of all polynomials $R(x_1, \cdots, x_n)$ satisfying certain conditions. But once it has been ascertained one can learn from the structure of this group a lot about the natural procedures by which to solve the equation. Galois' ideas, which for several decades remained a book with seven seals but later exerted a more and more profound influence upon the whole development of mathematics, are contained in a farewell letter written to a friend on the eve of his death, which he met in a silly duel at the age of twenty-one. This letter, if judged by the novelty and profundity of ideas it contains, is perhaps the most substantial piece of writing in the whole literature of mankind. I give two examples of Galois' theory.

The first is taken from antiquity. The ratio $\sqrt{2}$ between diagonal and side of a square is determined by the quadratic equation with rational coefficients

(2) $$x^2 - 2 = 0.$$

Its two roots are $\vartheta_1 = \sqrt{2}$ and $\vartheta_2 = -\vartheta_1 = -\sqrt{2}$,

$$x^2 - 2 = (x - \sqrt{2})(x + \sqrt{2}).$$

As I mentioned a moment ago, they are irrational. The deep impression which this discovery, ascribed to the school of Pythagoras, made on the thinkers of antiquity is evidenced by a number of passages in Plato's dialogues. It was this insight which forced the Greeks to couch the general doctrine of quantities in geometric rather than algebraic terms. Let $R(x_1, x_2)$ be a polynomial of x_1, x_2 with rational coefficients vanishing (i.e. assuming the value zero) for $x_1 = \vartheta_1$, $x_2 = \vartheta_2$. The question is whether $R(\vartheta_2, \vartheta_1)$ is also zero. If we can show that the answer is affirmative for every R then the transposition

$$(3) \qquad \vartheta_1 \rightarrow \vartheta_2, \qquad \vartheta_2 \rightarrow \vartheta_1$$

is an automorphism as well as the identity $\vartheta_1 \rightarrow \vartheta_1, \vartheta_2 \rightarrow \vartheta_2$. The proof runs as follows. The polynomial $R(x, -x)$ of one indeterminate x vanishes for $x = \vartheta_1$. Its division by $x^2 - 2$,

$$R(x, -x) = (x^2 - 2) \cdot Q(x) + (ax + b)$$

leaves a remainder $ax + b$ of degree 1 with rational coefficients a, b. Substitute ϑ_1 for x: the resulting equation $a\vartheta_1 + b = 0$ contradicts the irrational nature of $\vartheta_1 = \sqrt{2}$ unless $a = 0$, $b = 0$. Hence

$$R(x, -x) = (x^2 - 2) \cdot Q(x),$$

and consequently $R(\vartheta_2, \vartheta_1) = R(\vartheta_2, -\vartheta_2) = 0$. Thus the fact that the group of automorphisms contains the transposition (3) besides the identity is equivalent to the irrationality of $\sqrt{2}$.

My other example is Gauss' construction of the regular 17-gon with ruler and compass, which he found as a young lad of nineteen. Up to then he had vacillated between classical philology and mathematics; this success was instrumental in bringing about his final decision in favor of mathematics. In a plane we represent any complex number $z = x + yi$ by the point with the real Cartesian coordinates (x, y). The algebraic equation

$$z^p - 1 = 0$$

has p roots which form the vertices of a regular p-gon. $z = 1$ is one vertex; and since

$$(z^p - 1) = (z - 1) \cdot (z^{p-1} + z^{p-2} + \cdots + z + 1),$$

the others are the roots of the equation

(4) $\qquad z^{p-1} + \cdots + z + 1 = 0.$

If p is a prime number, as we shall now assume, they are algebraically indiscernible

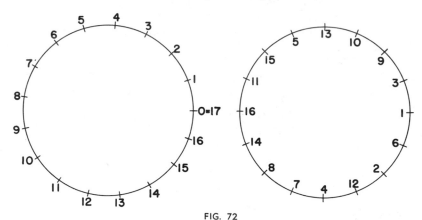

FIG. 72

and the group of automorphisms for the $p - 1$ roots is a cyclic group of order $p - 1$. I describe the situation for the case $p = 17$. The left 17-point dial (see Fig. 72) shows the labeling of the vertices, the right 16-point dial the 16 roots of (4) in a mysterious cyclic arrangement: the dialing of this diagram, i.e. iteration of its rotation by $\frac{1}{16}$ of the whole periphery, gives the 16 automorphisms as permutations among the 16 roots. This group C_{16} evidently has a subgroup C_8 of index 2; it is obtained by turning the dial through $\frac{1}{8}$, $\frac{2}{8}$, $\frac{3}{8}$, $\cdots$ of the full angle. By repeating this process of skipping alternate points we find a chain of consecutive subgroups ($\supset$ means "contains")

$$C_{16} \supset C_8 \supset C_4 \supset C_2 \supset C_1$$

which starts with the full group C_{16} and ends with the group C_1 consisting of the identity only, a chain in which each group is contained in the preceding one as a subgroup of index 2. Due to this circumstance one can determine the roots of the equation (4) by a chain of 4 consecutive equations of degree 2. Equations of degree 2, quadratic equations, are solved (as the Sumerians already knew) by extraction of square roots. Hence the solution of our problem requires, besides the rational operations of addition, subtraction, multiplication, and division, four consecutive extractions of square roots. However, the four species and extraction of a square root are exactly those algebraic operations which may geometrically be carried out by *ruler and compass*. This is the reason why the regular triangle, pentagon and 17-gon, $p = 3$, 5, and 17, may be constructed by ruler and compass; for in each of these cases the group of automorphisms is a cyclic group whose order $p - 1$ is a power of 2,

$$3 = 2^1 + 1, \quad 5 = 2^2 + 1, \quad 17 = 2^4 + 1.$$

It is amusing to observe that, whereas the (obvious) geometric symmetry of the 17-gon is described by a cyclic group of order 17, its (hidden) algebraic symmetry, which determines its constructibility, is described by one of order 16. It is certain that the regular heptagon is not constructible nor are the regular polygons with 11 and 13 sides.

Only if p is a prime number such that $p - 1$ is a power of 2, $p - 1 = 2^n$, then, according to Gauss's analysis, the regular p-gon is constructible by ruler and compass. However, $p = 2^n + 1$ cannot be a prime number unless the exponent n is a power of 2. For assume that 2^ν is the exact power of 2 by which n is divisible, so that $n = 2^\nu \cdot m$ where m is an odd number. Put $2^{2^\nu} = a$; then $2^n + 1 = a^m + 1$. But for odd m the number $a^m + 1$ is divisible by $a + 1$,

$$a^m + 1 =$$
$$(a + 1)(a^{m-1} - a^{m-2} + \cdots - a + 1),$$

and hence a composite number with the factor $a + 1$, unless $m = 1$. Therefore the next number of the form $2^n + 1$ after 3, 5, and 17 which has a chance to be a prime number is $2^8 + 1 = 257$. As this is actually a prime number, the regular 257-gon is constructible by ruler and compass.

Galois' theory may be put in a slightly different form, as I shall illustrate by the equation (2). Let us consider all numbers of the form $\alpha = a + b\sqrt{2}$ with rational components a, b; we call them the numbers of the field $\{\sqrt{2}\}$. Because of the irrationality of $\sqrt{2}$ such a number is zero only if $a = 0$, $b = 0$. Consequently the rational components a, b are uniquely determined by α, for $a + b\sqrt{2} = a_1 + b_1\sqrt{2}$ yields

$$(a - a_1) + (b - b_1)\sqrt{2} = 0;$$
$$a - a_1 = 0, \ b - b_1 = 0$$

or $a = a_1$, $b = b_1$, provided a, b and a_1, b_1 are

rational. Obviously addition, subtraction, and multiplication of two numbers of the field give rise to a number of the field. Nor does the operation of division lead beyond the field. For let $\alpha = a + b\sqrt{2}$ be a number of the field different from zero with the rational components a, b and let $\alpha' = a - b\sqrt{2}$ be its "conjugate." Because 2 is not the square of a rational number, the so-called norm of α, the rational number $\alpha\alpha' = a^2 - 2b^2$, is different from zero, and therefore one obtains the reciprocal $\dfrac{1}{\alpha}$ of α as a number in the field as follows:

$$\frac{1}{\alpha} = \frac{\alpha'}{\alpha\alpha'} = \frac{a - b\sqrt{2}}{a^2 - 2b^2}.$$

Thus the field $\{\sqrt{2}\}$ is closed with respect to the operations of addition, subtraction, multiplication, and division, with the self-understood exclusion of division by zero. We may now ask for the automorphisms of such a field. An automorphism would be a one-to-one mapping $\alpha \to \alpha^*$ of the numbers of the field such that $\alpha + \beta$ and $\alpha \cdot \beta$ go into $\alpha^* + \beta^*$ and $\alpha^* \cdot \beta^*$ respectively, for any numbers α, β in the field. It follows at once that an automorphism changes every rational number into itself and $\sqrt{2}$ into a number ϑ satisfying the equation $\vartheta^2 - 2 = 0$, thus either into $\sqrt{2}$ or $-\sqrt{2}$. Hence there are only two possible automorphisms, the one which carries every number α of the field $\{\sqrt{2}\}$ into itself, and the other carrying any number $\alpha = a + b\sqrt{2}$ into its conjugate $\alpha' = a - b\sqrt{2}$. It is evident that this second operation is an automorphism, and one has thus determined the group of all automorphisms for the field $\{\sqrt{2}\}$.

A field is perhaps the simplest algebraic

structure we can invent. Its elements are numbers. Characteristic for its structure are the operations of addition and multiplication. These operations satisfy certain axioms, among them those that guarantee a unique inversion of addition, called subtraction, and a unique inversion of multiplication (provided the multiplier is different from zero), called division. Space is another example of an entity endowed with a structure. Here the elements are points, and the structure is established in terms of certain basic relations between points such as: A, B, C lie on a straight line, AB is congruent CD, and the like. What we learn from our whole discussion and what has indeed become a guiding principle in modern mathematics is this lesson: *Whenever you have to do with a structure-endowed entity Σ try to determine its group of automorphisms*, the group of those element-wise transformations which leave all structural relations undisturbed. You can expect to gain a deep insight into the constitution of Σ in this way. After that you may start to investigate symmetric configurations of elements, i.e. configurations which are invariant under a certain subgroup of the group of all automorphisms; and it may be advisable, before looking for such configurations, to study the subgroups themselves, e.g. the subgroup of those automorphisms which leave one element fixed, or leave two distinct elements fixed, and investigate what discontinuous or finite subgroups there exist, and so forth.

In the study of groups of transformations one does well to stress the mere structure of such a group. This is accomplished by attaching arbitrary labels to its elements and then expressing in terms of these labels for any two group elements s, t what the result $u = st$

of their composition is. If the group is finite one could tabulate the composition of elements. The group scheme or abstract group thus obtained is itself a structural entity, its structure represented by the law or table of composition for its elements, $st = u$. Here the dog bites into its own tail, and maybe that is a clear enough warning for us to stop. Indeed one may ask with respect to a given abstract group: What is the group of its automorphisms, what are the one-to-one mappings $s \rightarrow s'$ of the group into itself which make st go over into $s't'$ while the arbitrary elements s, t go over into s', t' respectively?

Symmetry is a vast subject, significant in art and nature. Mathematics lies at its root, and it would be hard to find a better one on which to demonstrate the working of the mathematical intellect. I hope I have not completely failed in giving you an indication of its many ramifications, and in leading you up the ladder from intuitive concepts to abstract ideas.

APPENDICES

APPENDIX A

DETERMINATION OF ALL FINITE GROUPS OF PROPER ROTATIONS IN 3-SPACE (cf. p. 77).

A SIMPLE PROOF for the completeness of the list (5) in Lecture II is based on the fact first established by Leonhard Euler in the eighteenth century that every proper rotation in 3-space which is not the identity I is rotation around an axis, i.e. it leaves fixed not only the origin O but every point on a certain straight line through O, the axis l. It is sufficient to consider the two-dimensional sphere Σ of unit radius around O instead of the three-dimensional space; for every rotation carries Σ into itself and thus is a one-to-one mapping of Σ into itself. Every proper rotation $\neq I$ has two fixed points on Σ which are antipodes of each other, namely the points where the axis l pierces the sphere.

Given a finite group Γ of proper rotations of order N, we consider the fixed points of the $N - 1$ operations of Γ which are different from I. We call them poles. Each pole p has a definite multiplicity ν ($= 2$ or 3 or 4 or $\cdot\cdot\cdot$): The operations S of our group which leave p invariant consist of the iterations of the rotation around the corresponding axis by $360°/\nu$, and hence there are exactly ν such operations S. They form a cyclic subgroup Γ_p of order ν. One of these operations is the identity, hence the number of operations $\neq I$ leaving p fixed amounts to $\nu - 1$.

For any point p on the sphere we may consider the finite set C of those points q into which p is carried by the operations of the group; we call them points equivalent to p.

Because Γ is a group this equivalence is of the nature of an equality, i.e. the point p is equivalent to itself; if q is equivalent to p then p is equivalent to q; and if both q_1 and q_2 are equivalent to p then q_1 and q_2 are equivalent among each other. We speak of our set as a *class* of equivalent points; any point of the class may serve as its representative p inasmuch as the class contains with p all the points equivalent to p and no others. While the points of a sphere are indiscernible under the group of all proper rotations, the points of a class remain even indiscernible after this group has been limited to the finite subgroup Γ.

Of how many points does the class C_p of the points equivalent to p consist? The answer: of N points, that naturally suggests itself, is correct provided I is the only operation of the group which leaves p fixed. For then any two different operations S_1, S_2 of Γ carry p into two different points $q_1 = pS_1$, $q_2 = pS_2$ since their coincidence $q_1 = q_2$ would imply that the operation $S_1 S_2^{-1}$ carries p into itself, and would thus lead to $S_1 S_2^{-1} = I$, $S_1 = S_2$. But suppose now that p is a pole of multiplicity ν so that ν operations of the group carry p into itself. Then, I maintain, the number of points q of which the class C_p consists equals N/ν.

Proof: Since the points of the class are indiscernible even under the given group Γ, each must be of the same multiplicity ν. Let us first demonstrate this explicitly. If the operation L of Γ carries p into q then $L^{-1}SL$ carries q into q provided S carries p into p. Vice versa, if T is any operation of Γ carrying q into itself then $S = LTL^{-1}$ carries p into p and hence T is of the form $L^{-1}SL$ where S is an element of the group Γ_p.

Thus if $S_1 = I$, S_2, $\cdots$, S_ν are the ν elements leaving p fixed then

$$T_1 = L^{-1}S_1L, \qquad T_2 = L^{-1}S_2L, \cdots,$$
$$T_\nu = L^{-1}S_\nu L$$

are the ν different operations leaving q fixed. Moreover, the ν different operations S_1L, $\cdots$, $S_\nu L$ carry p into q. Vice versa, if U is an operation of Γ carrying p into q then UL^{-1} carries p into p and thus is one of the operations S leaving p fixed; therefore $U = SL$ where S is one of the ν operations S_1, $\cdots$, S_ν. Now let q_1, $\cdots$, q_n be the n different points of the class $C = C_p$ and let L_i be one of the operations in Γ carrying p into q_i ($i = 1, \cdots, n$). Then all the $n \cdot \nu$ operations of the table

$$S_1L_1, \cdots, S_\nu L_1,$$
$$S_1L_2, \cdots, S_\nu L_2,$$
$$\cdots \cdots \cdots \cdots$$
$$S_1L_n, \cdots, S_\nu L_n$$

are different from each other. Indeed each individual line consists of different operations. And all the operations of, say, the second line must be different from those in the fifth line since the former carry p into q_2 and the latter into the point $q_5 \neq q_2$. Moreover every operation of the group Γ is contained in the table because any one of them carries p into one of the points q_1, $\cdots$, q_n, say into q_i, and must therefore figure in the ith line of our table.

This proves the relation $N = n\nu$ and thus the fact that the multiplicity ν is a divisor of N. We use the notation $\nu = \nu_p$ for the multiplicity of a pole p; we know that it is the same for every pole p in a given class C, and it can therefore also be denoted in an unambiguous manner by ν_C. The multiplicity ν_C and the number n_C of poles in the class C are connected by the relation $n_C \nu_C = N$.

After these preparations let us now consider all pairs (S, p) consisting of an operation $S \neq I$ of the group Γ and a point p left fixed by S—or, what is the same, of any pole p and any operation $S \neq I$ of the group leaving p fixed. This double description indicates a double enumeration of those pairs. On the one hand there are $N - 1$ operations S in the group that are different from I, and each has two antipodic fixed points; hence the number of the pairs equals $2(N - 1)$. On the other hand, for each pole p there are $\nu_p - 1$ operations $\neq I$ in the group leaving p fixed, and hence the number of the pairs equals the sum

$$\sum_p (\nu_p - 1)$$

extending over all poles p. We collect the poles into classes C of equivalent poles and thus obtain the following basic equation:

$$2(N - 1) = \sum_C n_C(\nu_C - 1)$$

where the sum to the right extends over all classes C of poles. On taking the equation $n_C \nu_C = N$ into account, division by N yields the relation

$$2 - \frac{2}{N} = \sum_C \left(1 - \frac{1}{\nu_C}\right).$$

What follows is a discussion of this equation.

The most trivial case is the one in which the group Γ consists of the identity only; then $N = 1$, and there are no poles.

Leaving aside this trivial case we can say

that N is at least 2 and hence the left side of our equation is at least 1, but less than 2. The first fact makes it impossible for the sum to the right to consist of one term only. Hence there are at least two classes C. But certainly not more than 3. For as each ν_C is at least 2, the sum to the right would at least be 2 if it consisted of 4 or more terms. Consequently we have either two or three classes of equivalent poles (Cases II and III respectively).

II. In this case our equation gives

$$\frac{2}{N} = \frac{1}{\nu_1} + \frac{1}{\nu_2} \qquad \text{or} \qquad 2 = \frac{N}{\nu_1} + \frac{N}{\nu_2}.$$

But two positive integers $n_1 = N/\nu_1$, $n_2 = N/\nu_2$ can have the sum 2 only if each equals 1:

$$\nu_1 = \nu_2 = N; \qquad n_1 = n_2 = 1.$$

Hence each of the two classes of equivalent poles consists of *one* pole of multiplicity N. What we find here is the cyclic group of rotations around a (vertical) axis of order N.

III. In this case we have

$$\frac{1}{\nu_1} + \frac{1}{\nu_2} + \frac{1}{\nu_3} = 1 + \frac{2}{N}.$$

Arrange the multiplicities ν in ascending order, $\nu_1 \leq \nu_2 \leq \nu_3$. Not all three numbers ν_1, ν_2, ν_3 can be greater than 2; for then the left side would give a result that is $\leq \frac{1}{3} + \frac{1}{3} + \frac{1}{3} = 1$, contrary to the value of the right side. Hence $\nu_1 = 2$,

$$\frac{1}{\nu_2} + \frac{1}{\nu_3} = \frac{1}{2} + \frac{2}{N}.$$

Not both numbers ν_2, ν_3 can be ≥ 4, for then the left sum would be $\leq \frac{1}{2}$. Therefore $\nu_2 = 2$ or 3.

First alternative III_1: $\nu_1 = \nu_2 = 2$,
$$N = 2\nu_3.$$

Second alternative III_2: $\nu_1 = 2$, $\nu_2 = 3$;
$$\frac{1}{\nu_3} = \frac{1}{6} + \frac{2}{N}.$$

Set $\nu_3 = n$ in Case III_1. We have two classes of poles of multiplicity 2 each consisting of n poles, and one class consisting of two poles of multiplicity n. It is easily seen that these conditions are fulfilled by the dihedral group D_n' and by this group only.

For the second alternative III_2 we have, in view of $\nu_3 \geq \nu_2 = 3$, the following three possibilities:

$\nu_3 = 3, \quad N = 12; \qquad \nu_3 = 4, \quad N = 24;$
$\nu_3 = 5, \quad N = 60,$

which we denote by T, W, P respectively.

T: There are two classes of 4 three-poles each. It is clear that the poles of one class must form a regular tetrahedron and those of the other are their antipodes. We therefore obtain the tetrahedral group. The 6 equivalent two-poles are the projections from O onto the sphere of the centers of the 6 edges.

W: One class of 6 four-poles, forming the corners of a regular octahedron; hence the octahedral group. One class of 8 three-poles (corresponding to the centers of the sides); one class of 12 two-poles (corresponding to the centers of the edges).

Case P: One class of 12 five-poles which must form the corners of a regular icosahedron. The 20 three-poles correspond to the centers of the 20 sides, the 30 two-poles to the centers of the 30 edges of the polyhedron.

INCLUSION OF IMPROPER ROTATIONS (cf. p. 78).

If the finite group Γ^* of rotations in 3-space contains improper rotations let A be one of them and $S_1, \cdots, S_n$ be the proper operations in Γ^*. The latter form a subgroup Γ, and Γ^* contains one line of proper operations and another line of improper operations

$$\text{(1)} \qquad S_1, \cdots, S_n,$$
$$\text{(2)} \qquad AS_1, \cdots, AS_n.$$

It contains no other operations. For if T is an improper operation in Γ^* then $A^{-1}T$ is proper and hence identical with one of the operations in the first line, say S_i, and therefore $T = AS_i$. Consequently the order of Γ^* is $2n$, half of its operations are proper forming the group Γ, the other half are improper.

We now distinguish two cases according to whether the improper operation Z is or is not contained in Γ^*. In the first case we choose Z for A and thus get $\Gamma^* = \bar{\Gamma}$.

In the second case we may also write the second line in the form

$$\text{(2')} \qquad ZT_1, \cdots, ZT_n$$

where the T_i are proper rotations. But in this case all the T_i are different from all the S_i. Indeed were $T_i = S_k$ then the group Γ^* would contain with $ZT_i = ZS_k$ and S_k also the element $(ZS_k)S_k^{-1} = Z$, contrary to the hypothesis. Under these circumstances the operations

$$\text{(3)} \qquad \begin{matrix} S_1, \cdots, S_n \\ T_1, \cdots, T_n \end{matrix}$$

155

form a group Γ' of proper rotations of order $2n$ in which Γ is contained as a subgroup of index 2. Indeed the statement that the two lines (3) form a group is, as one easily verifies, equivalent to the other that the lines (1) and (2$'$) constitute a group (namely the group Γ^*). Thus Γ^* is what was denoted in the main text by $\Gamma'\Gamma$, and we have thus proved that the two methods mentioned there are the only ones by which finite groups containing improper rotations may be constructed.

ACKNOWLEDGMENTS

I wish to acknowledge especially my debt to Miss Helen Harris in the Marquand Library of Princeton University who helped me to find suitable photographs of many of the art objects pictured in this book. I am also grateful to the publishers who generously allowed me to reproduce illustrations from their publications. These publications are listed below.

Figs. 10, 11, 26.   Alinari photographs.

Fig. 15.   Anderson photograph.

Figs. 67, 68.   Dye, Daniel Sheets, *A grammar of Chinese lattice*, Figs. C9b, S12a. Harvard-Yenching Institute Monograph V.   Cambridge, 1937.

Figs. 69, 70, 71.   Ewald, P. P., *Kristalle und Röntgenstrahlen*, Figs. 44, 45, 125. Springer, Berlin, 1923.

Figs. 36, 37.   Haeckel, Ernst, *Kunstformen der Natur*, Pls. 10, 28.   Leipzig und Wien, 1899.

Fig. 45.   Haeckel, Ernst, Challenger monograph. *Report on the scientific results of the voyage of H.M.S. Challenger*, Vol. XVIII, Pl. 117.   H.M.S.O., 1887.

Fig. 54.   Hudnut Sales Co., Inc., advertisement in *Vogue*, February 1951.

Figs. 23, 24, 31.   Jones, Owen, *The grammar of ornament*, Pls. XVI, XVII, VI. Bernard Quaritch, London, 1868.

Fig. 46.   Kepler, Johannes, *Mysterium Cosmographicum*.   Tübingen, 1596.

Fig. 48.   Photograph by I. Kitrosser.   Réalités.   1er no., Paris, 1950.

Fig. 32.   Kühnel, Ernst, *Maurische Kunst*, Pl. 104.   Bruno Cassirer Verlag, Berlin, 1924.

Figs. 16, 18.   Ludwig, W., *Rechts-links-Problem im Tierreich und beim Menschen*, Figs. 81, 120a.   Springer, Berlin, 1932.

Fig. 17.   Needham, Joseph, *Order and life*, Fig. 5.   Yale University Press, New Haven, 1936.

Fig. 35.   New York Botanical Garden, photograph of Iris rosiflora.

Fig. 29.   Pfuhl, Ernst, *Malerei und Zeichnung der Griechen;* III. Band, Verzeichnisse und Abbildungen, Pl. I (Fig. 10).   F. Bruckmann, Munich, 1923.

Figs. 62, 65.   Speiser, A., *Theorie der Gruppen von endlicher Ordnung*, 3. Aufl., Figs. 40, 39.   Springer, Berlin, 1924.

Figs. 3, 4, 6, 7, 9, 25, 30.   Swindler, Mary H., *Ancient painting*, Figs. 91, (p. 45), 127, 192, 408, 125, 253.   Yale University Press, New Haven, 1929.

Figs. 42, 43, 44, 50, 51, 52, 55, 56.   Thompson, D'Arcy W., *On growth and form*, Figs. 368, 418, 448, 156, 189, 181, 322, 213.   New edition, Cambridge University Press, Cambridge and New York, 1948.

Fig. 53.   Reprinted from *Vogue Pattern Book*, Condé Nast Publications, 1951.

Figs. 27, 28, 39.   Troll, Wilhelm, "Symmetriebetrachtung in der Biologie," *Studium Generale*, 2. Jahrgang, Heft 4/5, Figs. (19 & 20), 1, 15.   Berlin-Göttingen-Heidelberg, Juli, 1949.

Fig. 38.   U.S. Weather Bureau photograph by W. A. Bentley.

Figs. 22, 58, 59, 60, 61, 64.   Weyl, Hermann,
"Symmetry," *Journal of the Washington
Academy of Sciences*, Vol. 28, No. 6,
June 15, 1938.   Figs. 2, 5, 6, 7, 8, 9.

Figs. 8, 12.   Wulff, O., *Altchristliche und
byzantinische Kunst;* II, Die byzantinische
Kunst, Figs. 523, 514.   Akademische
Verlagsgesellschaft Athenaion, Berlin,
1914.

INDEX

absolute vs. relative space and time, 21
abstract group, 145
active (optically active) substances, 17, 29
Adam and Eve, 32
addition of vectors, 93
affine geometry, 96, 132
algebra, 121, 135
Alhambra (Granada), 109, 112
analytic geometry, 94
Angraecum distichum, 51
a priori statements and symmetry, 126
Arabian Nights, 90
ARCHIBALD, R. C., 72
ARCHIMEDES, 92
ARISTOTLE, 3
arithmetical theory of quadratic forms, 120
arrangement of atoms in a crystal, 122
Ascaris megalocephala, 36
associative law, 95
asymmetry and symmetry in art, 13, 16; historic —,
    16; — of crystals, 17, 29; — in the animal kingdom,
    26; — the prerogative of life? 31; — and the origin
    of life, 32
aurea sectio, 72
automorphism, 18, 42; geometric and physical —,
    129–130; in algebra, 137; for any structure-
    endowed entity, 144
axis, 109, 149

BALMER series, 133
band ornaments, 48–51
bar, 52
beauty and symmetry, 3, 72
DE BEER, G. R., 34
bees, their geometric intelligence, 91
BEETHOVEN (pastoral sonata), 52
bilateral symmetry, Chap. I; definition, 4; — of
    animals, 26; aesthetic and vital value, 6, 28; of the
    human body, 8
bipotentiality of plasma, 38
BIRKHOFF, G. D., 3, 53
BONNET, CHARLES, 72
BROWNE, Sir THOMAS, 64, 103
BURIDAN's ass, 19

Capella Emiliana (Venice), 65
Cartesian coordinates, 97
causal structure, 25, 132

echinoderms, 60
EINSTEIN, ALBERT, 130, 132
enantiomorph crystals, 28
ENTEMENA, King —, 8
EPICURUS, 128
equiangular spiral, 69
equivalence of left and right, 19–20, 129; — of past and future, 24; — of positive and negative electricity, 25
equivalent points, 150
EUCLID, 17, 74
EUDOXUS, 136
EULER, L., 149; his topological formula ("polyhedron formula"), 89
exclusion principle, 134

FAISTAUER, A., 24
FAUST, 45
FEDOROW, 92
FIBONACCI series, 72
field of numbers, 143
flowers, their symmetry, 58
FONTENELLE, 91
FREY, DAGOBERT, 16
Fucus, 34
future, see: past

GALEN, 4
GALOIS, E., Galois group and Galois theory, 137, 142
GAUSS, 120, 139, 142
general relativity theory, 132
genetic constitution, 37
genotype and phenotype, 126
genotypical and phenotypical inversion, 37
geometry, what is a — ?, 133
geometric automorphism, 130
geranium, 66
gliding axis, 109
GOETHE, 51, 72
group of automorphisms, 42, 144; — of dilatations, 68; — of permutations, 135; — of transformations, 43; abstract group, 145; finite group of proper rotations in 2 dimensions, 54, — — of proper and improper rotations, 65; such groups having an invariant lattice, 120; the same for 3 dimensions, 79, 149–154; 80, 120, 155–156; 120; finite group of unimodular transformations in 2 dimensions, 107, in 3 dimensions, 120.

HAECKEL, ERNST, 60, 75, 88
HAMBIDGE, J., 72
HARRISON, R. G., 35
Helianthus maximus, 70, 73

helix, 71
helladic, 12
HELMHOLTZ, H., 18, 43, 128
heraldic symmetry, 8
HERMITE, 121
HERZFELD, ERNST, 10
hexagonal symmetry, 63; — lattice or pattern, 83, 110
HILBERT, D., 90
historic asymmetry, 16
HODLER, F., 16
homogeneous linear transformation, 96
Homo sapiens, 30
honey-comb, 83, 90–91
HUXLEY, JULIAN S., 34
hypercomplex number system, 121

icosahedron, 74
identity, 41
indiscernible, 17, 127
infinite rapport, 47
invariant quadratic form, 106, 108; — quantities, relations, etc., 133
inverse mapping, 41
inversion of left and right, 23; — of time, 24, 52; genotypical and phenotypical —, 37; situs inversus, 30
irrationality of square root of two, 139

JAEGER, F. M., 29
JAPP, F. R., 31
JONES, OWEN, 113
JORDAN, PASCUAL, 32

KANT, I., 21, 130
KELVIN, Lord, 93
KEPLER, 76, 90
KLEIN, FELIX, 133
KOENIG, SAMUEL, 91

laevo-, see under: dextro-
lattice, 100; — basis, 100; — structure and metric structure, 104; Chinese window —, 113; rectangular and diamond lattices, 102
LAUE, MAX VON, 122–124; LAUE interference pattern or diagram, 124
law of rational indices, 121
left and right: polar opposites or not?, 17, 22; concern a screw, 17; in paintings, 23
LEIBNIZ, 17, 18, 20, 21, 27, 127
LEONARDO DA VINCI, 66, 99
linear independence of vectors, 94; linear transformation, homogeneous, 96, and non-homogeneous, 97
logarithmic spiral, 69

longitudinal reflection, 50
LORENTZ, H. A., 131; LORENTZ group, 131
LORENZ, ALFRED, 52
LUDWIG, W., 26, 38

MACH, ERNST, 19
macroscopic and microscopic symmetry of crystals, 123
magnetism, positive and negative?, 20
Mainz cathedral, 56
MANN, THOMAS, 64
mapping, 18, 41
MARALDI, 90, 91
MASCHKE, H. (and MASCHKE's theorem), 107
Medusa, 66: medusae. 60
metric ground form, 96; — structure and lattice structure, 104
MICHELANGELO, 22
MINKOWSKI, H., 90, 121
mitosis, 34
modul, 96
Monreale, 14
morphology, 109, 126
motion (in the geometric sense), 44
multiplication of a vector by a number, 94
music, formal elements of —, 52
mythical thinking, 22

Nautilus, 70
NEEDHAM, JOSEPH, 34
negative, see under: positive
NEWTON, 20, 27, 43, 130
Nicomachean Ethics, 4
NIGGLI, PAUL, 122

objective statements, 128
octahedron, 74
ontogenesis of bilateral symmetry (and asymmetry), 33
OPPÉ, PAUL, 24
optically active, 17
order of a group, 79
ornamental symmetry in one dimension, 48; — — in two dimensions, 99–115
orthogonal transformation, 97
orthogonally equivalent, 99

parenchyma of maize, 87
particle (and wave), 25
past and future, 24, 132
PASTEUR, 29–31
PAULI, W., 134
Penicillium glaucum, 30

165

rotational symmetry, 5; — — of animals, 27; as
defined by a finite group of rotations, 53

San Apollinare (Ravenna), 13
San Marco (Venice), 14
S. Maria degli Angeli (Florence), 65
St. Matthew, 22
San Michele di Murano (Venice), 65
St. Pierre (Troyes, France), 58
seal stones, 10
sectio aurea, 72
SIEGEL, C. L., 121
similarity, 18, 42, 127
sinister, 22
Sistine Chapel, 22
situs inversus, 30
slip reflection, 50
space, 17; absolute or relative?, 21
space-time, 131
special relativity theory, 132
spectra, 133
SPEISER, ANDREAS, 50, 52, 74
spherical symmetry, 25, 27
spin, 134
'spiral' symmetry, 70
spira mirabilis, 69
Stephan's dome (Vienna), 67
structure of space, 17–18, 130; — of space-time, 131;
structure-endowed entity, 144; causal —, 25, 132
subgroup of index 2, 44
Sumerians, 8, 9
swastika, 66
symmetry = harmony of proportions, 3, 16; —
defined by a group of automorphisms, 45, 133, 134;
— and equilibrium, 25; — and a priori statements,
126; the three stages of spherical, rotational and
bilateral symmetry in the animal kingdom, 27;
symmetry of crystals, 121–123; symmetry classes of
crystals, 122; — of flowers, 58; bilateral, dilatory,
etc. symmetry, see under: bilateral, dilatory, etc.

TAIT, P. G., 73
tartaric acid, 29
tetrahedron, 74
tetrakaidekahedron, 92
theory of relativity, 127
THOMPSON, D'ARCY, 60
THOMSON, WILLIAM (Lord KELVIN), 93
time, absolute or relative?, 21; as the fourth dimen
sion of the world, 130
Tiryns, 12
Tomb of the Triclinium, 13
Torcello, 12